GC/WORKS/1 WITHOUT QUANTITIES (1998) GENERAL CONDITIONS

GW00724459

London: The Stationery Office

CONTENTS

SECURITY

MATERIALS AND WORKMANSHIP

COMMENCEMENT, PROGRAMME, DELAYS AND COMPLETION

INSTRUCTIONS AND PAYMENT

PARTICULAR POWERS AND REMEDIES

ASSIGNMENT, SUBLETTING, SUBCONTRACTING, SUPPLIERS AND OTHER WORKS

PERFORMANCE BOND, PARENT COMPANY GUARANTEE AND COLLATERAL WARRANTIES

INTRODUCTION AND NOTICES

THE GC/WORKS FAMILY OF CONTRACTS

GC/Works/1 (1998) is a new edition of the standard Government forms of contract for major UK building and civil engineering works, replacing the following three standard forms of contract -

- *GC/Works/1 (Edition 3) Lump Sum With Quantities, December 1989, revised 1990.*
- *GC/Works/1 (Edition 3) Lump Sum Without Quantities, August 1991.*
- *GC/Works/1 (Edition 3) Single Stage Design & Build Version, July 1993.*

GC/Works/1 (1998) is published in four volumes -

- With Quantities General Conditions.
- Without Quantities General Conditions.
- Single Stage Design & Build General Conditions.
- Model Forms & Commentary

PROCUREMENT METHODS

The With Quantities version is for use with Bills of Quantities, where all or most of the quantities are firm and not subject to re-measurement, giving a lump sum contract subject to adjustment for variations ordered.

The Without Quantities version is for use when lump sum tenders are to be invited on the basis of Specification and Drawings only, without Bills of Quantities, but supported by a Schedule of Rates prepared by the Contractor in order to value variations ordered.

The Single Stage Design & Build version is also a lump form of contract, which is intended to support a single stage tendering procedure, without a separate design stage. The form of contract is sufficiently flexible to allow for varying amounts of design input from the Contractor. The Employer provides Employer's Requirements, to which the Contractor responds with Contractor's Proposals, and the Contractor develops the detailed design on the basis outlined in those documents.

SUPPORTING DOCUMENTATION

Each of the three volumes of General Conditions also contains the following forms -

- Abstract of Particulars and Addendum.
- Invitation to Tender.
- Tender and Tender Price Form.
- Contract Agreement (England, Wales & Northern Ireland).
- Contract Agreement (Scotland).

The Model Forms & Commentary volume contains forms which may be used with any of the three sets of General Conditions.

DISCLAIMER

All parties must rely exclusively upon their own skill and judgement, or upon those of their advisers, when making use of this document. Neither the Crown, nor Pinsent Curtis, nor any other contributor, assumes any liability to anyone for any loss or damage caused by any error or omission, whether such error or omission is the result of negligence or any other cause. Any and all such liability is disclaimed.

SUPPLEMENTARY CONDITIONS

Conditions, such as a variation of price condition, may be added to supplement the printed Conditions of Contract according to circumstances. They are incorporated into the Conditions of Contract by listing in the Abstract of Particulars.

If the parties incorporate Supplementary Conditions into the Conditions of Contract by listing in the Abstract of Particulars, such Supplementary Conditions will, as provided by the Abstract, prevail over the printed Conditions.

GENERAL CONDITIONS OF CONTRACT GC/WORKS/1 (1998)

CONTRACT DOCUMENTATION, INFORMATION AND STAFF

1

Definitions, etc.

(1) In the Contract, unless the context otherwise requires -

'the Abstract of Particulars' means the document so headed included in the Contract;

'the Accepted Risks' means the risks of -

(a) pressure waves caused by the speed of aircraft or other aerial devices;

(b) ionising radiations or contamination by radioactivity from any nuclear fuel or from nuclear waste from the combustion of nuclear fuel;

(c) the radioactive, toxic, explosive or other hazardous properties of any explosive nuclear assembly (including any nuclear component); and

(d) war, invasion, act of foreign enemy, hostilities (whether or not war has been declared), civil war, rebellion, insurrection, or military or usurped power;

'CDM Regulations' means the Construction (Design and Management) Regulations 1994 or, if the Works are in Northern Ireland, the Construction (Design and Management) Regulations (Northern Ireland) 1995;

'Company' means and includes any body corporate;

'the Contract' means the -

(a) Contract Agreement (if any);

(b) Conditions of Contract (namely, these Conditions and any Supplementary Conditions and Annexes prescribed by the Abstract of Particulars);

(c) Abstract of Particulars;

(d) Specification;

(e) Drawings;

(f) Schedule of Rates;

(g) Programme;

(h) tender; and

(i) Employer's written acceptance;

'the Contract Agreement' means 'the formal agreement (if any) executed by the Employer and the Contractor recording the terms of the Contract;

'the Contract Sum' means the sum accepted by the Employer when awarding the Contract;

'the Contractor' means the person or persons whose tender is accepted by the Employer and his or their legal personal representatives or permitted assignees;

'the Date or Dates for Completion' means the date or dates ascertained in accordance with the Abstract of Particulars or, where extensions of time have been awarded or acceleration agreed, the date as adjusted by such extensions or agreements;

'Days' means calendar days;

'the Drawings' means the drawings listed in the Schedule of Drawings annexed to the Contract Agreement or, if there is no such Schedule, the drawings listed in the Schedule of Drawings referred to in the tender;

'the Employer' means the Employer in the Contract as named in the Abstract of Particulars and any permitted assignees;

'the Final Account' means the document prepared by the QS showing the calculation of the Final Sum and agreed or otherwise settled under Condition 49 (Final Account);

'the Final Sum' means the amount payable under the Contract by the Employer to the Contractor for the full and entire execution and completion of the Works;

'Group' means and includes a company and every holding company of that company for the time being, and every subsidiary for the time being of every such holding company, and the terms Employer's Group and Contractor's Group shall be interpreted accordingly; but, while the Employer is a Minister of the Crown, a government department or other Crown agency or authority, the term Employer's Group shall also include all other Ministers of the Crown, government departments and Crown agencies and authorities;

'the Health and Safety Plan' means, where it is stated in the Abstract of Particulars that all the CDM Regulations apply, the plan provided to the Principal Contractor and developed by him to comply with Regulation 15 of the CDM Regulations and, for the purpose of Regulation 10 of the CDM Regulations, received by the Employer before any construction work under the Contract has started; and any further development of that plan by the Principal Contractor during the progress of the Works;

'Holding company' shall have the meaning given in Section 736 of the Companies Act 1985, as substituted by Section 144 of the Companies Act 1989;

'Instruction' means any instruction given in accordance with Condition 40 (PM's Instructions), and, except where expressly stated to the contrary, includes any Variation Instruction;

'the Maintenance Period' means the period, or any of the periods, specified in the Abstract of Particulars for the rectification of defects in accordance with Condition 21 (Defects in Maintenance Periods);

'Milestone' means the completion of each of the phases of the Works described in the Milestone Payment Chart (if any);

'Milestone Payment Chart' means the chart or table (if any) included with the invitation to tender or agreed by the Employer which specifies the amounts of the advance payments to be made to the Contractor during the performance of the Works upon achievement of the Milestones;

'Planning Supervisor' means the PM, or any other person named in the Abstract of Particulars as Planning Supervisor, or such other person as may be appointed in that capacity for the time being by the Employer pursuant to Regulation 6(5) of the CDM Regulations;

'Principal Contractor' means the Contractor, or such other contractor or organisation as may be appointed as the Principal Contractor for the time being by the Employer pursuant to Regulation 6(5) of the CDM Regulations;

'the PM' means the Project Manager who is the person employed in that capacity named in the Abstract of Particulars and appointed by the Employer to act on his behalf in carrying out those duties described in the Contract (subject to the exclusions set out in the Abstract of Particulars), or such other person as may be appointed in that capacity for the time being by the Employer;

'the Programme' means the programme submitted prior to acceptance of the tender and agreed at that time by the Employer, as it may be amended from time to time;

'the QS' means the Quantity Surveyor appointed for the time being by the Employer;

'Retention Payment' shall have the meaning given in Condition 48A (Retention payment bond);

'the Schedule of Rates' means the schedule included in the Contract specifying the rates to be used for the purpose of valuing Variation Instructions;

'Section' means any part of the Works specified as a Section in the Abstract of Particulars and which has a specified Date for Completion;

'the Site' means the land or place described in the Contract, together with such other land or places as may be allotted or agreed by the parties from time to time, for the purpose of carrying out the Contract;

'the Specification' means the Specification annexed to the Contract Agreement or otherwise included in the Contract;

'the Stage Payment Chart' means the chart, table or graph (if any) included with the invitation to tender or agreed by the Employer which specifies the periods and the percentage amount of the advance payments to be made to the Contractor during the performance of the Works;

'Subsidiary' shall have the meaning given in Section 736 of the Companies Act 1985, as substituted by Section 144 of the Companies Act 1989;

'Things' comprise 'Things for incorporation', which means goods and materials intended to form part of the completed Works, and 'Things not for incorporation' which means goods and materials provided or used to facilitate execution of the Works but not for incorporation in them;

'Unforeseeable Ground Conditions' means ground conditions certified by the PM in accordance with Condition 7 (Conditions affecting Works);

'Variation Instruction' ('VI') means any Instruction which makes any alteration or addition to, or omission from, the Works or any change in the design, quality or quantity of the Works; and

'the Works' means the works described or shown in the Specification, Schedule of Rates and Drawings, including all modified or additional works to be executed under the Contract.

(2) The headings to these Conditions shall not affect their interpretation. Any reference to legislation shall be deemed to include a reference to any amendment or re-enactment thereof for the time being in force. Words in the singular include the plural, and *vice versa*. Words in the masculine include feminine and neuter.

(3) Unless otherwise provided by the Contract, certificates of the PM shall be issued to the Contractor. Any notices or other documents required or authorised to be served in pursuance of the Contract shall be in writing. They may be given to the Contractor by delivery to his agent or may be posted to the registered office or last known place of business of the Contractor. They may be given to the Employer by delivery to the PM or may be posted to the registered office or last known place of business of the Employer. A postal notice shall be deemed to have been served on the date when in the ordinary course of post it would have been delivered.

(4) (a) Except in relation to Condition 36 (Extensions of time), any period of time in these Conditions within which the Employer, the Contractor, PM or QS, is to take action or decide anything may be extended by agreement, notwithstanding that the period of time has expired.

(b) If the Works are in England, Wales or Scotland, then for the purposes of this Contract, and for all other purposes, periods of time shall include Saturdays and Sundays, but shall be reckoned as follows -

(i) where an act is required to be done within a specified period after or from a specified date, the period begins immediately after that date;
and

(ii) where the period would include Christmas Day, Good Friday or a day which under the Banking and Financial Dealings Act 1971 is a bank holiday in England and Wales, or as the case may be, in Scotland, that day shall be excluded.

(c) If the Works are in Northern Ireland, then for the purposes of this Contract, and for all other purposes, subject to Condition 50(4) (Certifying payments) periods of time shall include Saturdays and Sundays, but shall be reckoned as follows -

(i) where a period of time is expressed to begin on, or to be reckoned from, a particular day, that day shall not be included in the period;

(ii) where the time limited for the doing of anything expires or falls upon a Sunday or a public holiday, the time so limited shall extend to, and the thing may be done, on the first following day that is not a Sunday or a public holiday; and

(iii) the expression public holiday shall include Christmas Day, Good Friday, any bank holiday appointed by or under any statutory provision, and any day appointed for public thanksgiving or mourning.

1A

Fair dealing and teamworking

(1) The Employer and the Contractor shall deal fairly, in good faith and in mutual co- operation, with

one another, and the Contractor shall deal fairly, in good faith and in mutual co-operation, with all his subcontractors and suppliers.

(2) Both parties accept that a co-operative and open relationship is needed for success, and that teamwork will achieve this. The project team for this purpose shall include, but shall not be limited to, the PM and his representatives; the QS; the Contractor's agent; and major subcontractors and suppliers engaged on the Works from time to time.

(3) As soon as possible, the PM shall call a project team meeting and agree a programme of regular meetings with the Contractor. Either the PM or the Contractor may also call additional meetings of the team, and if the other agrees may invite any other person needed for an effective meeting. The PM and the Contractor shall use the meetings jointly to develop proposals for reducing costs, by solutions that will, so far as possible, be to the benefit of all affected by potential or actual problems. The meetings shall also consider the issues of advance warning of anything which might raise costs or harm final quality of the Works.

2

Contract documents

(1) In case of discrepancy between the Supplementary Conditions and Annexes prescribed by the Abstract of Particulars and these Conditions, the former shall prevail. In case of discrepancy between the Conditions of Contract and other documents forming part of the Contract, the Conditions of Contract shall prevail.

(2) The Specification shall take precedence over the specification content of the Drawings, unless the PM Instructs otherwise.

(3) The Contractor shall inform the PM of discrepancies between the Specification and the Drawings, or discrepancies between different parts of the Drawings or different parts of the Specification, which the Contractor discovers when handling the documents in order to prepare or execute the Works.

(4) Figured dimensions on all drawings shall be followed in preference to the scale.

(5) The PM shall provide free to the Contractor as soon as reasonably practicable a copy of each of the Drawings, the Specification, and any drawings issued during the progress of the Works, in a form which the PM considers suitable for reproduction. The Contractor shall keep one copy of all such drawings and of the Specification on the Site, and the PM or his representative shall have access to them at all reasonable times.

3

(Not used)

4

Delegations and representatives

(1) All decisions to be made by or on behalf of the Employer under the Contract shall be communicated to the Contractor by the PM. The PM shall be deemed to be authorised to act generally for the Employer, subject to any excluded matters set out in the Abstract of Particulars. In relation to those excluded matters, the person or persons authorised to act for the Employer are identified in the Abstract of Particulars, but decisions in respect of excluded matters shall nevertheless be communicated to the Contractor by the PM.

(2) The PM and QS may expressly delegate in writing to named representatives any of their powers and duties. Where a Clerk of Works or Resident Engineer is appointed he shall exercise the powers of the PM under Condition 31 (Quality) and such other powers as the PM may delegate to him.

(3) The appointment of representatives by the Employer, PM or QS shall not prevent them from subsequently exercising directly any of the powers and duties conferred under the Contract.

(4) The Contractor shall as soon as possible be notified of all powers and duties delegated, and of the names of representatives, and of any subsequent changes.

5

Contractor's agent

The Contractor shall employ a competent agent to supervise the execution of the Works. Except when required to attend at the office of the PM, or when reasonably absent from the Site for other reasons, the agent shall be in attendance at the Site during all working hours. When the agent is not in attendance at the Site, the Contractor will so notify the PM, stating the period of and reason for the absence, together with the name of the agent's authorised deputy.

6

Contractor's employees

(1) The PM may at any time require the Contractor immediately to cease to employ in connection with the Contract any person, including the Contractor's agent, whose continued employment is in the opinion of the PM undesirable. The Contractor shall replace any such person with a suitably qualified person.

(2) Other than for causes outside his control, the Contractor shall not make changes in personnel named in his tender in connection with the Contract without the prior approval of the PM.

GENERAL OBLIGATIONS

7

Conditions affecting Works

(1) The Contractor shall in relation to the Site be deemed to have satisfied himself as to -

   (a) the existing roads, railways and other means of communication with or access to it;

   (b) its contours and boundaries;

   (c) the risk of damage by reason of any work to any property adjacent to the Works and injury to occupiers of that property;

   (d) the nature of the soil and material (whether natural or otherwise) to be excavated;

   (e) the conditions under which the Works will have to be carried out, including precautions to prevent nuisance and pollution;

   (f) the supply of and conditions affecting labour necessary to carry out the Works;

   (g) the facilities for obtaining any Things whether or not for incorporation; and

   (h) any other matters or information affecting or likely to affect the execution of, or price tendered for, the Works.

(2) (Not used)

(3) If, during the execution of the Works, the Contractor becomes aware of ground conditions (excluding those caused by weather but including artificial obstructions) which he did not know of, and which he could not reasonably have foreseen having regard to any information which he had or ought reasonably to have ascertained, he shall, as a condition precedent to any right or remedy in respect of such conditions, by notice immediately -

   (i) inform the PM of those conditions; and
   (ii) state the measures which he proposes to take to deal with them.

(4) If the PM agrees that the ground conditions specified in a notice under paragraph (3) could not reasonably have been foreseen by the Contractor having regard to any information he should have had in accordance with that paragraph and paragraph (1), he shall certify those conditions to be Unforeseeable Ground Conditions. The PM shall notify the Contractor of his decision.

(5) If, after ground conditions specified in a notice under paragraph (3) have been or should have been certified as Unforeseeable Ground Conditions in accordance with this Condition, and as a result of such Unforeseeable Ground Conditions, the Contractor in executing the Works properly carries out or omits any work which he would not otherwise have carried out or omitted, then, without prejudice to any Instruction given by the PM, the value of the work carried out or omitted shall be ascertained in accordance with Condition 42 (Valuation of Variation Instructions) and the Contract Sum shall be increased or decreased accordingly.

(6) No claim by the Contractor for additional payment will be allowed because he has misunderstood or misinterpreted anything mentioned in paragraph (1). The Contractor shall not be released from

any risks or obligations imposed on, or undertaken by, him under the Contract for this reason, or because he did not or could not foresee any matter which might affect or have affected the execution of the Works.

8

Insurance

(1) A party required under this Condition to effect or maintain insurance is called in this Condition the insuring party but, while the Employer is a Minister of the Crown, a government department or other Crown agency or authority, the Employer shall be deemed not to be the insuring party.

(2) The Contractor shall by such existing or new policies as he sees fit effect and maintain from the time he takes possession of the Site or any part of the Site or from the time he commences the execution of the Works (if earlier) to the expiration of the last Maintenance Period to expire, employer's liability insurance in respect of persons in his employment, appropriate to the nature of the Works. Such insurance shall comply with the Employer's Liability (Compulsory Insurance) Act 1969 (or, if the Works are in Northern Ireland, the Employer's Liability (Defective Equipment and Compulsory Insurance) (Northern Ireland) Order 1972) and any subordinate legislation made thereunder, and shall be for the minimum amount of £10,000,000 (or such other minimum amount as may be stated in the Abstract of Particulars) for any one occurrence or series of occurrences arising out of one event.

Alternative A

(3) The Contractor shall by such existing or new policies as he sees fit effect and maintain for the same period -

(a) construction 'all risks' insurance in the joint names of the Employer and the Contractor against loss or damage to the Works and Things for which the Contractor is responsible under the terms of the Contract for the full reinstatement value thereof (including transit and off-Site risks) plus 15% (or such other percentage as may be stated in the Abstract of Particulars) for professional fees; and

(b) public liability insurance against legal liability for personal injury to any persons and loss or damage to property arising from or in connection with the Works, which is not covered by employer's liability insurance under paragraph (2) or by insurance against loss or damage to the Works and Things under subparagraph (a), for the minimum amount stated in the Abstract of Particulars, such public liability insurance to include a provision for indemnity to the Employer in respect of the Contractor's liability under Condition 19 (Loss or damage);

provided that the insurance which the Contractor is required to effect and maintain under this paragraph need not cover loss or damage caused by any Accepted Risk.

Alternative B

(3) The Contractor shall effect and maintain insurance in accordance with the Summary of Essential Insurance Requirements attached to the Abstract of Particulars.

Alternative C

(3) (a) While the Employer is a Minister of the Crown, a government department or other Crown agency or authority, the Employer shall not be required to effect or maintain any insurance, but shall assume the risks of loss or damage which should have been insured if paragraph (3)(b) applied, and, in the event of loss or damage arising, shall be responsible as if paragraph

3(b) applied and the Employer had failed to effect and maintain insurance as described therein, but the Employer shall not be responsible for any amounts in excess of amounts which should have been insured, or any liability authorised to be retained, or risks which would not have been insured or would have been excluded by the application of the terms, exceptions or conditions of any such insurance.

(b) While the Employer is not a Minister of the Crown, a government department or other Crown agency or authority, the Employer shall effect and maintain insurance in accordance with the Summary of Essential Insurance Requirements attached to the Abstract of Particulars, but shall not be responsible for any amounts in excess of amounts insured, or any retained liability, or risks not insured or excluded by the application of the terms, exceptions or conditions of any such insurance.

(4) The other party shall have the right to receive, on request, a copy of insurances required to be effected or maintained by the insuring party under this Condition. The insuring party shall within 21 Days of the acceptance of the tender, and also within 21 Days of any subsequent renewal or expiry date of relevant insurances, send to the other party a certificate in the form attached to the Abstract of Particulars from his insurer or broker attesting that insurance has been effected in accordance with the Contract.

(5) All insurances required to be effected or maintained by the insuring party under this Condition shall be with reputable insurers, to whom the other party has no reasonable objection, lawfully carrying on such insurance business in the United Kingdom, and upon customary and usual terms prevailing for the time being in the insurance market. The said terms and conditions shall not include any term or condition to the effect that any insured must discharge any liability before being entitled to recover from the insurers, or any other term or condition which might adversely affect the rights of any person to recover from the insurers pursuant to the Third Parties (Rights Against Insurers) Act 1930 or the Third Parties (Rights Against Insurers) Act (Northern Ireland) 1930.

(6) All insurances required to be effected or maintained under paragraph (3) (Alternatives B or C) (if applicable) -

(a) shall be in the joint names of the Employer, such other persons as the Employer may reasonably require (including, without limitation, the Employer's consultants, any persons who have entered or shall enter into an agreement for the provision of finance in connection with the Works, and any persons who have acquired or shall acquire any interest in or over the Works or any part thereof), the Contractor and all subcontractors; and

(b) shall extend to cover loss or damage which the Contractor is responsible for making good under Condition 21 (Defects in Maintenance Periods) or which occurs while the Contractor is making good defects in the Works under that Condition.

(7) If, without the approval of the other party, the insuring party fails to effect and maintain insurance he is required to effect and maintain under this Condition as described, or obtains a different policy of insurance, or fails to provide a copy of insurances or certificates in accordance with this Condition, the other party may, but is not required to, effect and maintain appropriate insurance cover and deduct the cost of doing so from any payment due to the insuring party under the Contract, or recover such sum from the insuring party as a debt.

(8) If -

(a) the Works involve the refurbishment, alteration or extension of existing structures; and/or

(b) a completed part within the meaning of Condition 37 (Early possession) is certified by the PM as having been completed in accordance with the Contract;

the Employer shall bear the risks of fire, lightning, explosion, storm, tempest, flood, bursting or overflowing of water tanks, apparatus or pipes, earthquake, aircraft or other aerial devices or articles dropped therefrom, riot and civil commotion (including terrorism) in respect of loss or damage to the existing structures and contents for which the Employer is responsible, and in respect of loss or damage to the completed part from the date of its certification, except where Condition 19 (Loss or damage) applies to the relevant loss or damage and to the extent that the Contractor is not entitled to reimbursement under Condition 19(5).

(9) For the avoidance of doubt, it is agreed that nothing in this Condition shall relieve the Contractor from any of his obligations and liabilities under the Contract.

8A

Professional indemnity insurance for design (only applicable if stated in Abstract of Particulars)

(1) The Contractor shall maintain professional indemnity insurance covering (*inter alia*) all liability hereunder in respect of defects or insufficiency in design, upon customary and usual terms and conditions prevailing for the time being in the insurance market, and with reputable insurers lawfully carrying on such insurance business in the United Kingdom (in an amount not less than that required by the Abstract of Particulars) for any one occurrence or series of occurrences arising out of any one event, for a period beginning now and ending 12 years (or such other period as is required by the Abstract of Particulars) after certification under Condition 39 (Certifying completion) of the completion of the Works or the last Section thereof in respect of which completion is certified, or the determination of the Contract for any reason whatsoever, including (without limitation) breach by the Employer, whichever is the earlier, provided always that such insurance is available at commercially reasonable rates. The said terms and conditions shall not include any term or condition to the effect that the Contractor must discharge any liability before being entitled to recover from the insurers, or any other term or condition which might adversely affect the rights of any person to recover from the insurers pursuant to the Third Parties (Rights Against Insurers) Act 1930 or the Third Parties (Rights Against Insurers) Act (Northern Ireland) 1930. The Contractor shall not, without the prior approval in writing of the Employer, settle or compromise with the insurers any claim which the Contractor may have against the insurers and which relates to a claim by the Employer against the Contractor, or by any act or omission lose or prejudice the Contractor's right to make or proceed with such a claim against the insurers.

(2) The Contractor shall immediately inform the Employer if such insurance ceases to be available at rates that the Contractor considers to be commercially reasonable. Any increased or additional premium required by insurers by reason of the Contractor's own claims record or other acts, omissions, matters or things particular to the Contractor shall be deemed to be within commercially reasonable rates.

(3) The Contractor shall fully co-operate with any measures reasonably required by the Employer, including (without limitation) completing any proposals for insurance and associated documents, maintaining such insurance at rates above commercially reasonable rates if the Employer undertakes in writing to reimburse the Contractor in respect of the net cost of such insurance to the Contractor above commercially reasonable rates or, if the Employer effects such insurance at rates at or above commercially reasonable rates, reimbursing the Employer in respect of what the net cost of such insurance to the Employer would have been at commercially reasonable rates.

(4)   As and when reasonably required to do so by the Employer, the Contractor shall produce for inspection documentary evidence (including, if required by the Employer, the originals of the relevant insurance document) that his professional indemnity insurance is being maintained.

(5)   The above obligations in respect of professional indemnity insurance shall continue notwithstanding determination of the Contract for any reason whatsoever, including (without limitation) breach by the Employer.

9

Setting out

(1)   Subject to any express provision in the Contract to the contrary, the PM shall provide such dimensional drawings, levels and other information as he considers reasonably necessary to enable the Contractor to set out the Works. The Contractor shall set out the Works accordingly, and shall provide all the instruments, profiles, templates and rods for that purpose. The Contractor shall be solely responsible for the correctness of the setting out.

(2)   The Contractor shall provide, fix and be responsible for the maintenance of all stakes, templates, profiles, level marks, points and any other setting out apparatus which is required. He shall take all necessary precautions to prevent their removal, alteration or disturbance and shall be liable for the consequences of their removal, alteration or disturbance and for their efficient reinstatement.

10

Design

(1)   Where the Contractor, either by himself or by means of any employee, agent, subcontractor or supplier, is required under the Contract to undertake the design of any part of the Works, he shall in accordance with the Contract or as Instructed by the PM submit to the PM for approval two copies (or such other number as is stated in the Abstract of Particulars) of a suitable drawing, design document or other suitable design information relating to that work, in the form and medium stated in the Specification, or Instructed by the PM. The Contractor shall not commence any work to which such drawing, design document or design information relates unless the design has been approved in writing by the PM, and the Contractor shall not alter that design without the further written approval of the PM. The approval of the PM shall not relieve the Contractor of any liability which he would otherwise have in respect of the design in accordance with paragraph (2).

Alternative A

(2)   The Contractor's liability to the Employer in respect of any defect or insufficiency in any design undertaken by the Contractor himself or by means of any employee, agent, subcontractor or supplier shall be the same as would have applied to an architect or other appropriate professional designer who had held himself out as competent to take on work for such design and who had acted independently under a separate contract with the Employer and supplied such a design for, or in connection with, works to be carried out and completed by a contractor not being the supplier of the design.

Alternative B

(2)   The Contractor warrants to the Employer that any Works designed by the Contractor, or by any employee, agent, subcontractor or supplier of his, will be fit for their purposes, as made known to the Contractor by the Contract.

(3)   The Contractor's liability under this Condition shall not be affected by any warranty that the Employer may obtain from any subcontractor.

(4)   The copyright in all documents provided by the Contractor or any subcontractor in connection with the Works shall remain vested in the copyright owner (or as may be otherwise provided by the relevant subcontract), but the Employer and its appointee shall have, and the Contractor warrants and shall procure that the Employer and its appointee shall have, a licence to copy and use such documents, and to reproduce the designs contained in them, for any purpose related to the Works including, without limitation, the construction, completion, maintenance, letting, promotion, advertisement, reinstatement, repair and extension of the Works. The Contractor shall, at any time and on one or more occasions, if the Employer so requests and undertakes in writing to pay the Contractor's reasonable copying charges, promptly supply the Employer with conveniently reproducible copies of all such documents.

11
Statutory notices and CDM Regulations

(1)   (a)   The Contractor shall give all notices required by any Act of Parliament or by any regulations or bye-laws made under any Act which may be required in connection with the Works. He shall pay any fees or charges required to be paid under any Act, regulations or bye-laws in respect of the Works and supply all drawings and plans required in connection with any notice.

     (b)   In this Condition the expression 'Act of Parliament' shall include an Act of the Parliament of Northern Ireland, a measure of the Northern Ireland Assembly and an Order in Council made under Section 1(3) of the Northern Ireland (Temporary Provisions) Act 1972 or under paragraph 1(1) of Schedule 1 to the Northern Ireland Act 1974; and the expressions 'regulation' and 'bye-laws' shall include subordinate legislation made under an Act of Parliament.

(2)   The Contractor shall obtain the consent, permission or licence of any statutory undertakers, or any adjoining owners whose services or land may be affected by, or whose consent is necessary in connection with, the Works. The Contractor shall pay any licence fee or charge required in connection with any consent or licence.

(3)   The Employer shall reimburse the amount of any fee or charge paid by the Contractor which he had properly incurred in accordance with paragraphs (1) or (2).

(4)   Where the Contractor is and remains the Principal Contractor, he will meet all the duties of a Principal Contractor as set out in the CDM Regulations and he shall liaise with the Planning Supervisor and ensure that any employee, agent, subcontractor or supplier of his complies with the CDM Regulations and co- operates with the Planning Supervisor so as to enable the Planning Supervisor to carry out his duties under the CDM Regulations.

(5)   Before commencing the Works, the Contractor shall submit to the PM the Health and Safety Plan containing the information required by Regulation 15 of the CDM Regulations and the Contractor shall not commence the Works until the PM has confirmed in writing to the Contractor that the Plan is of an appropriate standard. Such confirmation shall not relieve the Contractor of his liability under the Contract or the CDM Regulations. The Contractor shall notify the PM and the Planning Supervisor of all amendments to the Plan.

(6)   Where the Employer, necessarily in order to comply with the CDM Regulations, appoints a successor to the Contractor as Principal Contractor, the Contractor will meet all reasonable

costs to the Employer of any successor Principal Contractor. Notwithstanding Condition 36 (Extensions of time), no extension of time will be given by reason of the removal of the Contractor as Principal Contractor, and/or of the appointment of any successor Principal Contractor.

(7) Provided a reasonable period of written notice is given to the Contractor by the Planning Supervisor, the Contractor will provide, and will ensure that any subcontractor provides through him, such information as the Planning Supervisor needs for the preparation of a health and safety file as required by Regulations 14(d)(e) and (f) of the CDM Regulations. If the Contractor is not the Principal Contractor, such enquiries will be routed via the Principal Contractor.

12

Intellectual property rights

(1) The Contractor shall pay any royalty, licence fee or other expense for the supply or use of any patent, process, drawing, model, plan, invention or information used or necessary for, or in connection with, the Works.

(2) Where the use or supply of any patent, etc., in accordance with paragraph (1) gives rise to any claim or proceedings against the Employer, the Contractor shall reimburse the Employer any costs and expenses reasonably incurred by the Employer in respect of such claim or proceedings.

(3) The Employer shall reimburse the Contractor the amount of any royalty, etc., incurred in accordance with paragraph (1) which -

(a) was necessarily incurred in order to comply with a VI; and

(b) was not reasonably contemplated under the Contract.

13

Protection of Works

(1) The Contractor shall during the execution of the Works take all reasonable measures and precautions needed to take care of the Site and the Works, and shall have custody of all Things on the Site against loss or damage from fire and any other cause. The Contractor shall be solely responsible for and shall take all reasonable and proper steps for protecting, securing, lighting and watching all places on or about the Works and the Site which may be dangerous to his workpeople or to any other person.

(2) The Contractor shall comply with any statutory regulations (whether or not binding on the Crown) which govern the storage and use of all Things which are brought on to the Site in connection with the Works.

14

Nuisance and pollution

The Contractor shall take all reasonable precautions to prevent any nuisance or inconvenience to the owners, tenants or occupiers of any other property and to the general public, and shall secure the efficient protection of all streams and waterways against pollution.

15

Returns

The Contractor's agent shall provide the PM with a return, in the form the PM shall direct, of the number and description of workpeople and the plant employed each Day on the Works.

16
Foundations

The Contractor shall not lay any foundations until the excavations for them have been examined and approved by the PM. No such examination or approval shall relieve the Contractor of any liability.

17
Covering work

The Contractor shall give reasonable notice to the PM whenever any work or Thing for incorporation is intended to be covered with earth or otherwise. In default of such notice, the Contractor shall, if required by the PM, uncover the work or Thing at his own expense.

18
Measurement

(1) The Contractor's representative shall, when required on reasonable notice by the QS, attend at the Works to take jointly with the QS measurements of the work executed. These measurements and any differences in relation to them shall be recorded in the manner required by the QS.

(2) The Contractor shall without extra charge provide the appliances and other things necessary for measuring the work.

(3) If the Contractor's representative fails to attend when required in accordance with this Condition, the QS may proceed to take any measurements.

19
Loss or damage

(1) This Condition applies to any loss or damage which arises out of, or is in any way connected with, the execution or purported execution of the Contract.

(2) The Contractor shall without delay and at his own cost reinstate, replace or make good to the satisfaction of the Employer or, if the Employer agrees, compensate the Employer for, any loss or damage.

(3) Where a claim is made, or proceedings are brought against the Employer in respect of any loss or damage, the Contractor shall reimburse the Employer any costs or expenses which the Employer may reasonably incur in dealing with, or in settling, that claim or those proceedings.

(4) The Employer shall notify the Contactor as soon as possible of any claim made, or proceedings brought, against the Employer in respect of any loss or damage.

(5) The Employer shall reimburse the Contractor for any costs or expenses which the Contractor incurs in accordance with paragraphs (2) and (3) to the extent that the loss or damage is caused by -

(a) the neglect or default of the Employer or of any other contractor or agent of the Employer;

(b) any Accepted Risk or Unforeseeable Ground Conditions; or

(c) any other circumstances which are outside the control of the Contractor or of any of his subcontractors or suppliers and which could not have been reasonably contemplated under the Contract; provided that this subparagraph shall not apply where the loss or damage is loss or damage falling within subparagraph 6(c).

(6)   In this Condition loss or damage includes -

(a)   loss or damage to property;

(b)   personal injury to, or the sickness or death of, any person;

(c)   loss or damage to the Works or to any Things on the Site; and

(d)   loss of profits or loss of use suffered because of any loss or damage.

20

(Not used)

21

Defects in Maintenance Periods

(1)   The Contractor shall without delay make good at his own cost any defects in the Works, resulting from what the Employer considers to be default by the Contractor or his agent or subcontractors or suppliers, which appear during the relevant Maintenance Period.

(2)   After completion of the remedial works by the Contractor, the Employer shall reimburse the Contractor for any cost the Contractor has incurred to the extent that the Contractor demonstrates that any defects were not caused by -

(a)   the Contractor's neglect or default, or the neglect or default of any agent or subcontractor of his; or

(b)   by any circumstances within his or their control.

(3)   If the Contractor fails to comply with this Condition, the Employer may do anything necessary to make good any defects notified to the Contractor. All the costs and expenses reasonably incurred by the Employer in doing so shall be recoverable from the Contractor.

(4)   In the case of any defects which have been made good under this Condition, the relevant Maintenance Period shall apply to the remedial works in full from the date of making good.

22

Occupier's rules and regulations (only applicable if stated in Abstract of Particulars)

The Contractor shall comply with the occupier's rules and regulations which have been provided to him or made available to him for inspection, both in respect of the Site and in respect of any larger premises of which the Site forms part. The Contractor shall comply with any changes to those rules and regulations notified to him as an Instruction under Condition 40 (PM's Instructions) during the execution of the Works.

23

Discrimination

(1)   The Contractor shall not unlawfully discriminate within the meaning and scope of the provisions of the Race Relations Act 1976, the Sex Discrimination Acts 1975 and 1986 or the Sex

Discrimination (Northern Ireland) Orders 1976 and 1988; and shall, if the Works are in Northern Ireland, conform with the requirements of the Fair Employment (Northern Ireland) Acts 1976 and 1989.

(2) The Contractor shall take all reasonable steps to ensure the observance of the provisions of paragraph (1) by all his employees, agents and subcontractors.

24

Corruption

(1) The Contractor shall not by himself, or in conjunction with any other person -

(a) corruptly solicit, receive or agree to receive, for himself or for any other person; or

(b) offer or agree to give to any person in the Employer's service, or any consultant or contractor who has a contract with the Employer;

any gift or consideration of any kind as an inducement or reward for doing or not doing anything, or for showing favour or disfavour to any person, in relation to this Contract or any other contract to which the Employer is a party.

(2) The Contractor shall not enter into this or any other contract with the Employer in connection with which commission has been paid or agreed to be paid by him or on his behalf or to his knowledge unless, before any such contract is made, particulars of any such commission, and of the terms and conditions of any agreement for the payment thereof, have been disclosed in writing to the Employer.

(3) The Employer may by notice determine the Contract if -

(a) he is reasonably satisfied that the Contractor or anyone employed by him or acting on his behalf (whether with or without the knowledge of the Contractor) is in breach of this Condition; or

(b) the Contractor or anyone employed by him or acting on his behalf is convicted of any offence under the Prevention of Corruption Acts 1889 to 1916 in relation to this Contract or any other contract to which the Employer is a party.

(4) If the Employer so determines the Contract, then (without prejudice to any powers conferred by Condition 51 (Recovery of sums) the Employer shall be entitled to recover from the Contractor the amount or value of any such gift, consideration or commission.

(5) In this Condition, all references to the Employer or the Contractor shall be deemed to include a reference to each member of their respective Groups.

25

Records

(1) The Contractor shall for the purposes of the Contract keep such records as may be reasonably necessary for the QS, the PM or the Employer to ascertain or verify any claims made or to be made by the Contractor or any sums to be paid to the Contractor under or in connection with the Contract.

(2) In order that the PM and QS may discharge their respective functions under the Contract, the Contractor shall afford them access to the records mentioned in paragraph (1) and supply them with the information (including means to interpret the records) they may require.

SECURITY

26

Site admittance

(1) The Contractor shall take the steps reasonably required by the PM to prevent unauthorised persons being admitted to the Site. If the PM gives the Contractor notice that any person is not to be admitted to the Site, the Contractor shall take all reasonable steps to prevent that person being admitted.

(2) If and when Instructed by the PM, the Contractor shall give to the PM a list of names and addresses of all persons who are or may be at any time concerned with the Works or any part of them, specifying the capacities in which they are so concerned, and giving such other particulars as the PM may reasonably require.

(3) The Contractor shall bear the cost of any notice, Instruction or decision of the PM under this Condition.

27

Passes (only applicable if stated in Abstract of Particulars)

Where passes are required for admission to the Site, the PM shall issue them to the Contractor. The Contractor shall submit to the Employer for his approval a list of the names of the workpeople and any other information the Employer reasonably requires for this purpose. The passes shall be returned at any time on the demand of the Employer and in any case on the completion of the Works.

28

Photographs (only applicable if stated in Abstract of Particulars)

The Contractor shall not at any time take any photograph of the Site or the Works or any part of them, and shall take all reasonable steps to ensure that no such photographs shall at any time be taken or published or otherwise circulated by any employee, agent or subcontractor of his, unless the Contractor has obtained the prior written consent of the PM.

29

Official secrets and confidentiality

(1) The Contractor shall take all reasonable steps to ensure that all persons employed by him or his subcontractors in connection with the Contract are aware of the Official Secrets Act 1989 and, where appropriate, of the provisions of Section 11 of the Atomic Energy Act 1946, and that these Acts apply to them during the execution of the Works and after the completion of the Works or earlier determination of the Contract.

(2) Any information concerning the Contract obtained either by the Contractor or by any person employed by him in connection with the Contract is confidential and shall not be used or disclosed by the Contractor or by any such person except for the purposes of the Contract.

MATERIALS AND WORKMANSHIP

30
Vesting

(1) The Works and any Things on the Site in connection with the Contract, the ownership of which the Contractor is able to transfer, or which vest in him under any contract, shall become the property of and vest in the Employer.

(2) Subject to Conditions 19 (Loss or damage) and 65 (Other works), the Employer shall not be responsible or chargeable for any Thing lost, stolen, damaged, destroyed or removed from the Site, or which is in any way unfit or unsuitable for its purpose.

(3) The Contractor shall be responsible for the protection and preservation of the Works and any Things brought on the Site until the completion of the Works or the determination of the Contract.

(4) No Things shall be removed from the Site before completion of the Works without the written consent of the PM. The PM may Instruct or permit the Contractor in writing at any time to remove from the Site any Things which are unused or which have been rejected by the PM, and the Contractor shall at his own expense forthwith remove them. Once so removed from the Site, the Things shall re-vest in the Contractor.

(5) In this Condition, all references to the Employer or the Contractor shall be deemed to include a reference to each member of their respective Groups.

31
Quality

(1) The Contractor shall execute the Works in accordance with the Contract and -

   (a) with diligence;

   (b) in accordance with the Programme;

   (c) with all reasonable skill and care; and

   (d) in a workmanlike manner.

(2) The Contractor warrants that all Things for incorporation, with the sole exception of Things for incorporation chosen or selected by the Employer by means of a statement by or on behalf of the Employer in the Contract or in a VI, shall be fit for their intended purposes, and shall conform to the requirements of the Specification, the Schedule of Rates and the Drawings.

(3) The Contractor shall notify the PM before incorporation of any Things that the Contractor considers should not be incorporated.

(4) The Contractor shall when requested by the PM demonstrate that he is performing his duties under paragraphs (1) and (2). The PM shall have power at any time to inspect and examine any part of the Works or inspect, examine and test any Things for incorporation either on the Site, or at any factory or workshop or other place where any such Thing is being constructed or manufactured, or at any place where it is lying, or from which it is being obtained. The Contractor shall give the PM the assistance and facilities he may reasonably require for any such inspection and examination. The PM may reject any Thing for incorporation which does not conform with

the Specification, Schedule of Rates or Drawings, or which, even though conforming, is not of good quality or is not fit for its purpose.

(5) The PM may arrange for an independent expert to test whether any Thing for incorporation is fit for use in the Works, of good quality, and conforms to the requirements of the Contract. The reasonable costs incurred by the Employer in arranging for an expert to carry out any test shall be borne by the Contractor, if the test results disclose that the Thing tested does not conform with the provisions of the Contract. The Contractor shall also bear the reasonable cost of any further tests reasonably required to monitor quality, following negative test results.

(6) The Contractor shall, at his own cost, replace, rectify or reconstruct -

(a) the Works, or any part of them, which do not conform with the Contract; and

(b) any Things for incorporation which do not conform with the Contract, and which have been rejected by the PM.

(7) If the Contract requires that the Contractor shall be accredited in accordance with any quality control or assurance scheme or system, the Contractor warrants that he is so accredited, and undertakes that he will continue to be so accredited until after the PM has issued a certificate under Condition 39 (Certifying completion) when the Contractor has complied with Condition 21 (Defects in Maintenance Periods). If the Contractor ceases for any reason to be so accredited during that period, he shall by notice immediately inform the PM, giving full particulars of the circumstances, and of his proposals to reinstate the relevant accreditation, or to obtain an alternative accreditation.

32
Excavations

(1) Except as otherwise provided by the Contract, material and objects of any kind obtained from work on the Site (including, without limitation, from excavations, demolition or dismantling) shall remain or become the property of the Employer.

(2) When the Employer's property is permitted to be used in substitution for any Things (whether or not for incorporation), which the Contractor would otherwise have provided, the QS shall ascertain the amount of any saving in the cost of the execution of the Works. The Contract Sum shall be reduced by the amount of any saving.

(3) All objects which are, or appear to be, fossils, antiquities, or likely to have interest or value, found on the Site or in carrying out excavations in the execution of the Works, shall remain or become the property of the Employer. Upon the discovery of any such object, the Contractor shall forthwith -

(a) take all practicable measures not to disturb the object;

(b) cease work, if the continuance of work would endanger, or disturb, the object, or prevent or impede its excavation or removal;

(c) take all necessary steps to preserve the object in the exact position and condition in which it was found; and

(d) inform the PM of the discovery and precise location of the object.

(4) Any Instructions issued by the PM in relation to any object mentioned in paragraph (3), may require the Contractor to permit the examination, excavation or removal of the object by a third party.

COMMENCEMENT, PROGRAMME, DELAYS AND COMPLETION

33

Programme

(1) The Contractor warrants that the Programme shows the sequence in which the Contractor proposes to execute the Works, details of any temporary work, method of work, labour and plant proposed to be employed, and events, which, in his opinion, are critical to the satisfactory completion of the Works; that the Programme is achievable, conforms with the requirements of the Contract, permits effective monitoring of progress, and allows reasonable periods of time for the provision of information required from the Employer; and that the Programme is based on a period for the execution of the Works to the Date or Dates for Completion.

(2) Subject to Conditions 35 (Progress meetings), 37 (Early possession) and 38 (Acceleration and cost savings), the Contractor may at any time submit for the PM s agreement proposals for the amendment of the Programme. The agreement of the PM to any proposal for the amendment of the Programme shall not relieve the Contractor of any liability which he has under the Contract. In particular, without limitation, the submission by the Contractor of any proposal for the amendment of the Programme showing a period for the execution of the Works extending beyond the Date or any of the Dates for Completion shall not constitute a notice from the Contractor requesting an extension of time for the completion of the Works or of any Section; and the agreement of the PM to any such amendment shall not constitute, or be evidence of, or in support of, any extension of time for the completion of the Works or of any Section.

34

Commencement and completion

(1) Within the period specified in the Abstract of Particulars, the Employer shall notify the Contractor when he may take possession of the Site, or those parts of the Site defined in the Contract. The date so notified shall be not more than 14 Days after the end of that period. The Date or Dates for Completion shall be calculated either from the date so notified, or from the acceptance of the tender, whichever is provided by the Abstract of Particulars. The Contractor shall accordingly take possession of the Site, or such parts of the Site defined in the Contract. The Contractor shall, subject to Condition 11 (Statutory notices and CDM Regulations), proceed with diligence and in accordance with the Programme or as may be Instructed by the PM, so that the whole of the Works or any relevant Section shall be completed in accordance with the Contract by the Date or Dates for Completion.

(2) The Contractor shall, at all times, keep the Site tidy and free from débris, litter, and rubbish and shall, not later than the completion of the Works, remove from the Site all Things for incorporation in the Works or any relevant Section which are unused, together with all Things not for incorporation. The Contractor shall, by the due date, clear and remove all rubbish and deliver up the Site and the Works in all respects to the satisfaction of the PM. The Contractor shall comply at his own cost with any Instructions relating to the removal of any Things and rubbish.

35

Progress meetings

(1) The Contractor's agent shall attend regular progress meetings to assess the progress of the

Works or any relevant Section, and to facilitate their due and satisfactory completion by the Date or Dates for Completion.

(2) An initial progress meeting shall be held, and thereafter progress meetings shall be held each month, unless Instructed to the contrary. The PM shall specify the time and place of progress meetings.

(3) The Contractor shall submit to the PM, 5 Days before each progress meeting, a written report which shall -

(a) describe the progress of the Works by reference to the Programme and relevant Instructions;

(b) specify all outstanding requests by the Contractor for drawings, nominations, levels or other information;

(c) explain any new circumstances, arising since any previous progress meeting, which in his opinion have delayed, or may delay, completion of the Works or a Section, or may increase the cost to the Employer of the Works or a Section, estimating any increase in such cost;

(d) refer to any request for an extension of time under Condition 36 (Extensions of time) since the previous progress meeting; and

(e) set out any re-programming proposals to ensure that completion of the Works or any Section will be achieved by the relevant Date for Completion.

(4) The PM shall, within 7 Days after each progress meeting, give the Contractor a written statement which specifies -

(a) by reference to the Programme the extent to which he considers the Works are on time, delayed or early;

(b) the matters which the PM considers have delayed, or are likely to delay, due completion of the Works or Section, or may increase the cost to the Employer of the Works or a Section, estimating any increase in such cost;

(c) the steps which the PM has agreed with the Contractor to reduce or eliminate the effects of any such delay, or to reduce to eliminate any estimated increase in the cost to the Employer of the Works or a Section;

(d) the situation in respect of applications for, and awards of, extensions of time under Condition 36 (Extensions of time); and

(e) his response to outstanding requests for drawings, nominations, levels or other information.

36

Extensions of time

(1) Where the PM receives notice requesting an extension of time from the Contractor (which shall include the grounds for his request), or where the PM considers that there has been or is likely to be a delay which will prevent or has prevented completion of the Works or any Section by the relevant Date for Completion (in this Condition called 'delay'), he shall, as soon as possible and in any event within 42 Days from the date any such notice is received, notify the Contractor of his decision regarding an extension of time for completion of the Works or relevant Section.

(2) The PM shall award an extension of time under paragraph (1) only if he is satisfied that the delay, or likely delay, is or will be due to -

(a) the execution of any modified or additional work;

(b) any act, neglect or default of the Employer, the PM or any other person for whom the Employer is responsible (not arising because of any default or neglect by the Contractor or by any employee, agent or subcontractor of his);

(c) any strike or industrial action which prevents or delays the execution of the Works, and which is outside the control of the Contractor or any of his subcontractors;

(d) an Accepted Risk or Unforeseeable Ground Conditions;

(e) any other circumstances (not arising because of any default or neglect by the Contractor or by any employee, agent or subcontractor of his, and other than weather conditions), which are outside the control of the Contractor or any of his subcontractors, and which could not have been reasonably contemplated under the Contract;

(f) failure of the Planning Supervisor to carry out his duties under the CDM Regulations properly; or

(g) the exercise by the Contractor of his rights under Condition 52 (Suspension for non-payment).

(3) The PM shall indicate whether his decision is interim or final. The PM shall keep all interim decisions under review until he is satisfied from the information available to him that he can give a final decision.

(4) No requests for extensions of time may be submitted after completion of the Works. The PM shall in any event come to a final decision on all outstanding and interim extensions of time within 42 Days after completion of the Works. The PM shall not be entitled in a final decision to withdraw or reduce any interim extension of time already awarded, except to take account of any authorised omission from the Works or any relevant Section that he has not already allowed for in an interim decision.

(5) The Contractor within 14 Days from receipt of a decision, submit a claim to the PM specifying the grounds which in his view entitle him to an extension or further extension of time. The PM shall by notice give his decision on a claim within 28 Days of its receipt.

(6) The Contractor must endeavour to prevent delays and to minimise unavoidable delays, and to do all that may be required to proceed with the Works. The Contractor shall not be entitled to an extension of time where the delay or likely delay is, or would be, attributable to the negligence, default, improper conduct or lack of endeavour of the Contractor.

37

Early possession

(1) The Employer shall be entitled, before the completion of the Works, to take possession of any part of the Works (in this Condition referred to as a completed part) which is certified by the PM as having been completed in accordance with the Contract and is either -

(a) a Section; or

(b) any other part of the Works in respect of which the parties agree, or the PM has given an Instruction, that possession shall be given before the completion of the Works or the relevant Section;

and the completed part, on and after the date on which the certificate is given, shall no longer form part of the Works for the purposes of Conditions 19 (Loss or damage) and 30 (Vesting).

(2) The provisions of Condition 21 (Defects in Maintenance Periods) shall have effect in relation to a completed part as if the Maintenance Period or Periods in respect of the completed part, or any subcontract works comprised in it, commenced on the date of certification under paragraph (1).

(3) As soon as possible after certification under paragraph (1), the PM shall certify the value of the completed part and his estimates for the purposes of paragraphs (4) and (5).

(4) Condition 55 (Liquidated damages) shall have effect notwithstanding that the Employer has taken possession of a completed part. Where the completed part comprises part of the Works as mentioned in subparagraph (1)(b), the rate of liquidated damages specified in the Abstract of Particulars in respect of the Works or the relevant Section shall be reduced by the same ratio as the value of the completed part bears to what the value of the Works or of the relevant Section will be when completed in accordance with the Contract, as estimated by the PM.

(5) The retention accumulated in accordance with Condition 48 (Advances on account) shall be apportioned by the PM with effect from the date of the certificate given under paragraph (1), so that the share of the retention apportioned in respect of a completed part shall bear the same ratio to the whole of the retention as the value of the completed part bears to the value of the Works which at that date have been completed in accordance with the Contract, as estimated by the PM.

(6) The Employer shall pay to the Contractor -

(a) one half of the share apportioned in accordance with paragraph (5) in respect of the completed part; and

(b) the remaining one half of that share, when the PM has certified after the appropriate Maintenance Period that in respect of the completed part the Contractor has complied with Condition 21 (Defects in Maintenance Periods).

38

Acceleration and cost savings

(1) If the Employer wishes to achieve completion of the Works or any Section before the Date or Dates for Completion, he shall direct the Contractor to submit to him within the period specified in the direction -

(a) the Contractor's priced proposals for achieving the accelerated completion date, together with any consequential amendments to the Programme; or

(b) the Contractor's explanation why he is unable to achieve the accelerated completion date.

(2) If the Employer accepts the Contractor's proposals he shall specify -

(a) the accelerated Date for Completion of the Works or any relevant Section;

(b) the amendments to the Programme, including any relevant critical paths and any supporting documentation;

(c) the amount by which the Contract Sum shall be adjusted;

(d) a revised Milestone or Stage Payment Chart or Charts, if necessary; and

(e) any other relevant amendment to the Contract which has been agreed with the Contractor.

(3) The Contractor may at any time submit to the Employer proposals for completing the Works or any Section before the Date or Dates for Completion. The Employer undertakes to consider any such proposals and if he accepts them to take action as stated in paragraph (2).

(4) The Contractor may, at any time during the carrying out of the Works, submit to the PM a written proposal which, in the Contractor's opinion, will enhance the buildability of the Works, or reduce the cost of the Works, or the cost of maintenance, or increase the efficiency of the completed Works. Any proposal shall clearly state that it is submitted for consideration under this Condition and shall include an estimate, for consideration by the Employer, of the amount to which the Contractor may be entitled on the basis that he and the Employer shall share equally the relevant savings as determined in accordance with Conditions 41 to 43 (Valuation of Instructions).

(5) The Contractor shall provide any further information relating to his proposal which either the PM or the Employer may require.

(6) The Contractor, having submitted such a proposal to the PM for consideration, shall continue with the expeditious carrying out of the Works.

(7) If the Employer accepts any such proposal, or agrees with the Contractor any amended proposal -

(a) the PM shall issue an Instruction to that effect;

(b) the Date or Dates for Completion, and the Programme, shall be amended accordingly; and

(c) the PM shall issue any agreed extension of time which is necessary.

38A

Bonuses (only applicable if stated in Abstract of Particulars)

If a rate of bonus for early completion of the Works or a Section has been specified in the Abstract of Particulars, and the Works or the relevant Section are or is completed before the relevant Date for Completion, the Employer shall pay the Contractor bonus at the rate specified in the Abstract of Particulars for each Day falling between the date certified by the PM under Condition 39 (Certifying completion) as the date of completion of the Works, or of the relevant Section, and the relevant Date for Completion.

39

Certifying completion

The PM shall certify the date when the Works, or any Section, or any completed part within the meaning of Condition 37 (Early possession), are completed in accordance with the Contract. Such completion shall include sufficient compliance by the Contractor with Condition 11(7) (Statutory notices and CDM Regulations). After the end of the last Maintenance Period to expire, he shall issue a certificate when the Contractor has complied with Condition 21 (Defects in Maintenance Periods).

INSTRUCTIONS AND PAYMENT

40

PM's Instructions

(1) The PM may from time to time issue further drawings, details, instructions, directions and explanations, all or any of which shall be treated for the purposes of the Contract as Instructions, including Variation Instructions. The Contractor shall comply forthwith with any Instruction.

(2) Instructions may be given in relation to all or any of the following matters -

    (a) the variation or modification of all or any of the Specification, Drawings or Schedule of Rates, or the design, quality or quantity of the Works;

    (b) any discrepancy in or between the Specification, Drawings and Schedule of Rates;

    (c) the removal from the Site of any Things for incorporation and their substitution with any other Things;

    (d) the removal and/or re-execution of any work executed by the Contractor;

    (e) the order of execution of the Works or any part of them;

    (f) the hours of working and the extent of overtime or night work to be adopted;

    (g) the suspension of the execution of the Works or any part of them;

    (h) the replacement of any person employed in connection with the Contract;

    (i) the opening up for inspection of any work covered up;

    (j) the amending and making good of any defects under Condition 21 (Defects in Maintenance Periods);

    (k) cost savings under Condition 38 (Acceleration and cost savings);

    (l) the execution of any emergency work as mentioned in Condition 54 (Emergency work);

    (m) the use or disposal of material obtained from excavations, demolition or dismantling on the Site;

    (n) the actions to be taken following discovery of fossils, antiquities or objects of interest or value;

    (o) measures to avoid nuisance or pollution;

    (p) quality control accreditation of the Contractor as mentioned in Condition 31 (Quality); and

    (q) any other matter which the PM considers necessary or expedient.

(3) All Instructions shall be in writing except those under subparagraphs (b), (d), (g) and (l) of paragraph (2), which may be given orally. Oral Instructions shall be immediately effective in accordance with their terms, but shall be confirmed in writing by the PM within 7 Days. Written Instructions shall be given to the Contractor's agent. Oral Instructions may be given to such

employee or agent of the Contractor as the PM thinks fit, but shall be confirmed in writing to the Contractor's agent. The Contractor's agent shall immediately acknowledge receipt of any written Instruction, and of every written confirmation of an oral Instruction.

(4) The Contractor shall not add to, omit from, or otherwise alter the Works except in accordance with an Instruction.

(5) The PM may include in a VI a requirement for the Contractor to submit to the QS not later than 21 Days from the receipt of that Instruction a written quotation of the lump sum total price for complying with it. The PM may make any VI conditional upon agreement of such a lump sum price, pending which agreement the Contractor is not to begin complying with the VI.

41

Valuation of Instructions - Principles

(1) The value (if any) of any VI shall be determined in accordance with Condition 42 (Valuation of Variation Instructions) and the value (if any) of any other Instruction in accordance with Condition 43 (Valuation of other Instructions).

(2) The value of any Instruction shall include any disruption to or prolongation of both varied and unvaried work.

(3) The value of any Instruction shall be added to, or as the case may be deducted from, the Contract Sum, except where the Instruction was necessary because of any default or neglect by the Contractor or by any employee, agent or subcontractor of his.

(4) The Contractor shall supply the QS with any information required by the QS to enable him to value a VI or determine the expense (if any) of complying with any other Instruction.

42

Valuation of Variation Instructions

(1) The value of a VI shall be determined -

    (a) by acceptance by the PM of a lump sum quotation prepared by the Contractor and submitted to the QS in accordance with Condition 40(5) (PM's Instructions) or such other lump sum as may be agreed by negotiation on such a quotation; or

    (b) by valuation by the QS in accordance with paragraphs (5) to (11).

(2) Any lump sum quotation submitted by the Contractor to the QS in accordance with Condition 40(5) (PM's Instructions) shall indicate how the lump sum was calculated by showing separately the amounts attributable to -

    (a) complying with that Instruction; and

    (b) any disruption to or prolongation of varied and unvaried work consequential upon compliance with the VI.

The Contractor shall include with his quotation such other information as will enable the QS to evaluate that quotation.

(3) The PM shall notify the Contractor, not later than 21 Days from the receipt of any such lump sum quotation, whether or not it is accepted or, if it is not acceptable, whether he is prepared to agree

any other lump sum. If accepted, the aggregate amount specified in that quotation, or otherwise agreed between the Employer and the Contractor, shall be the full sum to which the Contractor is entitled for complying with that VI.

(4) Should -

    (a) the Contractor fail to provide a lump sum quotation in accordance with Condition 40(5) (PM's Instructions); or

    (b) the Employer and the Contractor fail to agree by negotiation, and no agreement for independent assessment is reached;

then the PM shall instruct the QS to value the VI.

(5) Where the QS is required to value a VI he shall do so -

    (a) by measurement and valuation at the rates and prices in the Schedule of Rates for similar work; or

    (b) if it is not possible to value as in subparagraph (a), then by measurement and valuation at rates and prices deduced or extrapolated from the rates and prices in such Schedule; or

    (c) if it is not possible to value as in subparagraph (b), then by measurement and valuation at fair rates and prices, having regard to current market prices; or

    (d) if it is not possible to value by any of the preceding methods of measurement and valuation, then by the value of the materials used and plant and labour employed in accordance with the basis of charge for daywork described in the Contract.

(6) If, in the opinion of the QS, the VI prolongs and/or disrupts work not within the direct scope of the VI, then the QS shall adjust the rates for such work as he considers appropriate.

(7) The Contractor shall submit the information mentioned in Condition 41(4) (Valuation of Instructions - Principles) not later than 14 Days after its being requested by the QS.

(8) The QS shall, not later than 28 Days from the receipt of the information mentioned in paragraph (7), notify the Contractor of his valuation of the VI.

(9) If the Contractor disagrees with the whole or part of the QS s valuation he shall, within 14 Days of the QS s notification under paragraph (8), give his reasons for disagreement and his own valuation. In any other case he shall be treated as having accepted the notification under paragraph (8), and no further claim shall be made by him in respect of the VI.

(10) Any percentage or lump sum adjustments made in the pricing of the Schedule of Rates shall be deemed to be applicable to the pricing of all relevant VI s.

(11) Where an alteration in, or addition to, the Works would otherwise fall to be valued using rates and prices in the Schedule of Rates in accordance with subparagraphs (5)(a) or (b), but the QS is of the opinion that the relevant VI was issued at such a time or was of such content as to make it unreasonable for the alteration or addition to be so valued, he shall ascertain the value by measurement and valuation at fair rates and prices, having regard to current market prices.

(12) Where the QS has determined that work shall be valued in accordance with subparagraph (5)(d), the Contractor shall give reasonable notice to the QS before commencing the work, and shall deliver to him, by the end of the week following that in which the work was done, vouchers in

the form required by the QS, which specify separately the labour, materials and plant for that week.

(13) Where, as a result of a VI, the Contractor makes a saving in the cost of executing the Works, the Contract Sum shall be decreased by the amount of the saving as determined by the QS.

43

Valuation of other Instructions

(1) Where as the result of an Instruction, not being a VI, the Contractor -

    (a) properly and directly incurs any expense beyond that provided for in, or reasonably contemplated by, the Contract; or

    (b) makes any saving in the cost of executing the Works;

    the Contract Sum shall, subject to Condition 41(3) (Valuation of Instructions - Principles), be increased by the amount of the expense, or decreased by the amount of the saving, in either case as determined by the QS.

(2) The Contractor shall submit the information mentioned in Condition 41(4) (Valuation of Instructions - Principles) within 28 Days of complying with the relevant Instruction. Within 28 Days of receiving such information, the QS shall notify the Contractor of the amount he has determined.

(3) If the Contractor disagrees with the amount notified under paragraph (2), he shall within 14 Days of receipt notify the QS of his reasons for disagreement and of his own estimate of the correct amount. If the Contractor does not so notify, he shall be regarded as having accepted the amount determined by the QS, and may make no further claim in respect of the Instruction.

(4) In this Condition expense shall mean money expended by the Contractor, but shall not include any sum expended, or loss incurred, by him by way of interest or finance charges however described.

44

(Not used)

45

VAT

(1) All sums payable by or to the Employer or the Contractor are exclusive of Value Added Tax ('VAT'). Where VAT is chargeable on such sums, the payer shall pay, upon production of a valid VAT invoice by the payee, such VAT in addition to such sums.

(2) If the Contractor fails to carry out his obligations under the Contract and the Employer employs some other contractor to fulfil them, and a payment in respect of VAT is made or falls to be made to that other contractor, then the Employer shall be entitled to recover from the Contractor any VAT (which he is not otherwise able to recover) additional to what he would have paid had the Contractor carried out his obligations under the Contract.

(3) Where a party is liable to reimburse or indemnify the other party for costs incurred by that other party, the amount to be paid shall not include any VAT charged on such costs, save where the payee is unable to recover such VAT from HM Customs & Excise as input tax.

46

Prolongation and disruption

(1) If the Contractor properly and directly incurs any expense which he would not otherwise have incurred by reason of -

    (a) the execution of works pursuant to Condition 65 (Other works); or

    (b) any delay in being given possession of the Site or part of it, or any delay in respect of any of the matters specified in paragraph (2); or

    (c) any advice from the Planning Supervisor other than that properly required by the CDM Regulations;

which unavoidably results in the regular progress of the Works or of any part of them being materially disrupted or prolonged and which is beyond that provided for or reasonably contemplated by the Contract, and is not a consequence of any default or neglect on the part of the Contractor, or by any employee, agent or subcontractor of his, the Contract Sum shall be increased by the amount of that expense as determined by the QS.

(2) The matters referred to in subparagraph (1)(b) are -

    (a) any decisions, confirmations, consents, drawings, schedules, levels or other design information to be provided by the PM;

    (b) the execution of any work or the supply of any Thing by the Employer or ordered from somebody other than the Contractor; and

    (c) any direction or Instruction from the Employer or the PM regarding the issue of any pass to any person (if required under Condition 27 (Passes)) or any direction, Instruction or consent of the Employer or the PM to be given under Condition 63(2) (Nomination), provided that the Employer and the PM shall be entitled to a reasonable time for consideration and decision in respect of the matters specified in this subparagraph, and this Condition shall not fetter their proper discretion under those Conditions.

(3) The Contract Sum shall not be increased under paragraph (1) unless -

    (a) the Contractor, immediately upon becoming aware that the regular progress of the Works or any part of them has been or is likely to be disrupted or prolonged, has given notice to the PM specifying the circumstances causing or expected to cause that disruption or prolongation and stating that he is, or expects to be, entitled to an increase in the Contract Sum under that paragraph; and

    (b) the Contractor, as soon as reasonably practicable, and in any case within 56 Days of incurring the expense, provides to the QS full details of all expenses incurred and evidence that the expenses directly result from the occurrence of one of the events described in paragraph (1).

(4) Subject to paragraph (3), the Contract Sum shall be increased in accordance with subparagraph (1)(b) only where the Employer has failed to supply an item or act -

    (a) by a date agreed beforehand with the Contractor; or

    (b) within any reasonable period specified in a notice given by the Contractor to the Employer or the PM for the supply of the item or taking the action.

(5) The QS shall, not later than 28 Days from the receipt of the information referred to in subparagraph (3)(b), notify the Contractor of his decision under this Condition.

(6) In this Condition expense shall mean money expended by the Contractor, but shall not include any sum expended, or loss incurred, by him by way of interest or finance charges however described.

47

Finance charges

(1) The Employer shall pay the Contractor an amount by way of interest or finance charges (hereafter together called finance charges) only in the event that money is withheld from him under the Contract because either -

(a) the Employer, PM or QS has failed to comply with any time limit specified in the Contract or, where the parties agree at any time to vary any such time limit, that time limit as varied; or

(b) the QS varies any decision of his which he has previously notified to the Contractor.

(2) Finance charges shall be calculated as a percentage of the amounts which would have been paid to the Contractor if any of the events mentioned in paragraph (1) had not occurred. The rate at which finance charges shall be payable shall be the percentage stated in the Abstract of Particulars over the rate charged during the relevant period by the Bank of England for lending money to the clearing banks, compounded with effect from 31 March, 30 June, 30 September and 31 December.

(3) Finance charges shall be payable in respect of any period commencing with the date on which, but for a failure or variation mentioned in paragraph (1), money properly due under the Contract should have been certified, and ending with the date on which it was certified for payment under Condition 50 (Certifying payments).

(4) When calculating finance charges the QS shall take into account any overpayment made to the Contractor as a result of circumstances described in subparagraph (1)(b).

(5) The Employer shall not be liable to pay any finance charges which result from -

(a) any act, neglect or default of the Contractor or any of his subcontractors;

(b) any failure by the Contractor or any of his subcontractors to supply the PM or the QS with any relevant information; or

(c) any disagreement about the Final Account.

(6) The Employer and the Contractor agree that, when they entered into the Contract, neither of them had knowledge of any special facts or circumstances which would entitle the Contractor to be paid interest or finance charges, except in the circumstances mentioned in paragraph (1).

(7) The respective powers of the adjudicator and arbitrator to award interest shall be in addition to, and not in derogation from, this Condition.

48

Advances on account

(1) The Contractor shall following certification under Condition 50 (Certifying payments) be entitled

to be paid advances during the execution of the Works.

(2) The amount of each advance to be certified shall, subject to Condition 48A (Retention payment bond), be the total of the following sums -

Alternative A (Stage Payment Chart)

(a) 95% of the proportion of the sum specified as the proportion of the Contract Sum payable for the relevant month according to the Stage Payment Chart or Charts.

Alternative B (Milestone Payment Chart)

(a) Unless the Milestone Payment Chart provides otherwise, 95% of the cumulative value (not previously paid) of the Works according to the Milestone Payment Chart at the relevant Milestone achieved. Unless the Milestone Payment Chart provides otherwise, and specifies what payment is to be made upon the achievement of each Milestone, the Contractor shall not be entitled to any payment upon achievement of a Milestone until he has also achieved all the preceding Milestones set out in the Milestone Payment Chart. The Contractor shall submit applications in writing to the PM for payment under this subparagraph, upon the achievement of each Milestone. Unless and until the Contractor submits such an application, he shall not be entitled to certification or payment in respect of the relevant sum. If the Abstract of Particulars so specifies, such an application shall also include all amounts due under subparagraphs (2)(b)-(h), and the Contractor shall not be entitled to certification or payment of an advance in respect of such amounts except as included in such application. Paragraph (3) shall not apply.

Alternative C (Valuation)

(a) (i) 95% of the value of the work executed on the Site (other than such work referred to in subparagraphs (c) and/or (d)); and 95% of the value of any Things for incorporation which have been reasonably delivered to the Site and are adequately stored and protected against damage by weather and other causes, but which have not been incorporated in the Works.

(ii) Where any Things for incorporation on account of which an advance has been made under this subparagraph are incorporated in the Works, the amount of such advance may be deducted from the next or any subsequent payment made under the Contract.

(iii) The Contractor shall submit applications in writing to the PM for payment of advances on account of work executed and Things for incorporation which have been delivered to the Site, supported by a valuation of such work and Things, not later than 7 Days before such certificate is due under Condition 50 (Certifying payments). Unless and until the Contractor timely submits such an application and valuation, he shall not be entitled to certification or payment of an advance in respect of the relevant work on Things. Paragraph (3) shall not apply.

(b) 100% of any amount calculated under Condition 38A (Bonuses) (if applicable).

(c) 100% of any amount agreed under Condition 42(1)(a) (Valuation of Variation Instructions) in respect of work completed in the relevant month.

(d) 100% of the agreed value or, failing agreement, 95% of the QS's valuation, under Condition 42(5) (Valuation of Variation Instructions) and Condition 43 (Valuation of other Instructions) in respect of work completed in the relevant month.

    (e)   100% of any amount determined by the QS under Condition 46 (Prolongation and disruption) in respect of the relevant month.

    (f)   100% of any amount calculated under Condition 47 (Finance charges).

    (g)   95% of any amount calculated under Condition 48C (Payment for Things off-Site) (if applicable).

    (h)   Less, in each case, any sum agreed to be credited by the Contractor for old materials.

(3)   Where the PM has recorded in a statement after a progress meeting that the Works are in delay or are ahead of Programme he shall by reference to the Stage Payment Chart or Charts adjust the Contractor's entitlement to payments in accordance with paragraph (2)(a).

(4)   Before the payment of any advance or the issue of the final certificate for payment the Contractor shall, if requested by the PM, demonstrate to him that any amount due to a subcontractor or supplier of Things for incorporation which is covered by any previous advance has been paid. In any case where the PM is not satisfied the Employer may withhold payment to the Contractor of the amount in question until the PM is satisfied.

(5)   The Employer shall accumulate as a retention the balance of any sum withheld under paragraph (2) which is less than 100% and shall continue to hold the entire beneficial interest therein.

48A

Retention payment bond

If the Abstract of Particulars provides for the Employer to pay the Contractor advances on account without deduction of retention, provided that the amount of retention so foregone by the Employer shall not exceed the amount stated in the Abstract of Particulars ('the Retention Payment'), the Contractor shall be entitled to be paid advances on account without deduction of retention, after fulfilling the following conditions precedent -

    (a)   delivery to the Employer of a retention payment bond in the form prescribed by the Contract from the surety or sureties named in the tender in the amount of the Retention Payment;

    (b)   such delivery to occur within 28 Days of the acceptance of the tender, in respect of which period time shall be of the essence.

48B

Mobilisation payment (only applicable if stated in Abstract of Particulars)

(1)   If the Abstract of Particulars provides for the Employer to pay the Contractor a mobilisation payment, the Contractor shall be entitled to be paid that payment within 14 Days of fulfilling the following conditions precedent -

    (a)   delivery to the Employer of a mobilisation payment bond in the form prescribed by the Contract from the surety or sureties named in the tender in the amount of the mobilisation payment;

    (b)   such delivery to occur within 28 Days of the acceptance of the tender, in respect of which period time shall be of the essence.

(2)   The Employer shall be entitled to recover the mobilisation payment by deduction (in addition to any retention) of the percentage stated in the Abstract of Particulars either -

(a)   from the proportion of the sum specified as the proportion of the Contract Sum payable for the relevant month according to the Stage Payment Chart or Charts; or

(b)   from the cumulative value of the Works according to the Milestone Payment Chart at the relevant Milestone achieved in the relevant month; or

(c)   from the cumulative value of the work executed on Site and of Things for incorporation delivered to the Site;

whichever is applicable under Condition 48 (Advances on account).

(3)   Any part of the mobilisation payment not previously recovered by the Employer shall be immediately repayable by the Contractor to the Employer upon either -

(a)   the certification by the PM under Condition 39 (Certifying completion) of the completion of the Works or the last Section thereof in respect of which completion is certified; or

(b)   the determination of the Contract for any reason whatsoever, including (without limitation) breach by the Employer;

whichever occurs first.

48C

Payment for Things off-Site (only applicable if stated in Abstract of Particulars, *and not applicable in Scotland*)

(1)   Where this Condition applies compulsorily, as specified in the Abstract of Particulars, the PM shall, and where this Condition applies voluntarily, as so specified, the PM may in his complete discretion, include in his certificates under Condition 50 (Certifying payments) 95% of his estimate of the value at that time (ascertained on the basis of fair and reasonable prices) of Things to which this Condition applies, and which have been vested in the Employer in accordance with this Condition. For the purposes of this Condition, 'Things' shall mean any Things for incorporation which are manufactured, assembled or constructed off the Site and which the PM considers are in accordance with the Contract and substantially ready for incorporation in the Works. Any sum so certified and paid may, when the relevant Things have been delivered to the Site, or re-vest in the Contractor for any reason whatsoever, be deducted from any subsequent advance, without prejudice and in addition to any other rights and remedies of the Employer.

(2)   In order to transfer the property in Things, the Contractor shall -

(a)   ensure that the Things have been properly and securely set aside at the factory or workshop or other place where any such Things have been manufactured, assembled or constructed or at any place where they are lying or from which they are being obtained;

(b)   ensure that the Things have been suitably marked or otherwise identified so as to show that their destination is the Site, that they are the property of the Employer and, where relevant, to whose order they are held; and

(c)   provide to the PM documentary evidence that the property in the Things has vested unconditionally in the Contractor.

(3) Upon the PM confirming that the Contractor has complied with the requirements of paragraph (2), the Contractor shall transfer to the Employer the property in the Things. Upon the PM approving that transfer, the Things shall vest in and become the absolute property of the Employer. The Things shall thenceforth be in the possession of the Contractor for the sole purpose of the performance of the Contract and delivering the completed Things to the Site for inclusion in the Works, and shall not be within the ownership or disposition of the Contractor.

(4) Approval by the PM for the purposes of this Condition shall be without prejudice to the power of the PM to reject any Things in accordance with Condition 31 (Quality). In the event of any Things being rejected by the PM in accordance with Condition 31 (Quality), they shall re-vest in the Contractor.

(5) The Contractor shall be solely responsible for, and shall take all responsible and proper steps for protecting, preserving and securing any Things held off the Site.

(6) The Contractor shall comply with any statutory regulations (whether or not binding on the Crown) which govern the storage and use of all Things off the Site.

(7) The Contractor shall indemnify the Employer against all claims and proceedings made or brought against the Employer in respect of any loss and/or damage which arises out of, or is any way connected with, the manufacture, assembly, construction, storage or transportation of any Things. For the purpose of this Condition, loss or damage includes -

  (a) loss or damage to property;

  (b) personal injury to, or the sickness or death of, any person;

  (c) loss or damage to any Things; and

  (d) loss of profits or loss of use suffered because of any loss or damage.

(8) If the Contract is determined the Employer may give to the Contractor, before the expiration of 28 Days from the date on which such determination takes effect, one or more notices stating that the Employer elects that all or any Things which have not been delivered to the Site shall re-vest in the Contractor. The Things specified in the notice shall so re-vest upon service of that notice.

(9) With regard to Things which the Employer does not elect shall re-vest in the Contractor under paragraph (8) -

  (a) the Contractor shall hand over to the Employer the Things, and if he fails to do so, the Employer may enter any premises and remove the Things and recover the cost of doing so from the Contractor; and

  (b) subject to the terms of the Contract (in particular, without limitation, Condition 57 (Consequences of determination by Employer), if applicable to the determination of the Contract) the Employer shall pay a fair and reasonable price for the Things which are handed over to him by the Contractor or otherwise come into his possession, but shall be given credit for any payment therefor previously made under this Condition or otherwise.

(10) Any payment made by the Employer in respect of any Things which re-vest in the Contractor under paragraphs (4) or (8) shall be a sum recoverable in accordance with Condition 51 (Recovery of sums).

(11) The Contractor shall incorporate provisions equivalent to those provided in this Condition

in every subcontract in which provision is to be made for the manufacture, assembly or construction of Things off the Site.

49

Final Account

(1) Upon completion of the Works in accordance with the Contract the Employer shall as soon as reasonably possible pay to the Contractor an amount equal to the difference between -

   (a) the amount estimated by the Employer to be the Final Sum less one half the retention accumulated under Condition 48(5) (Advances on account); and

   (b) the total amount of advances paid under Condition 48 (Advances on account).

(2) Within 6 months of the certified completion of the Works or the last part thereof under Condition 39 (Certifying completion), the QS shall forward one copy of the draft final account to the Contractor. If the draft final account shows a Final Sum greater than the Final Sum estimated by the Employer under paragraph (1)(a), the Employer shall correct any payment made under paragraph (1) as soon as possible, in accordance with the Final Sum shown in the draft final account. The Contractor shall within 3 months of receipt of the draft final account notify his agreement or disagreement with the draft final account. The Contractor shall specify in any notice of disagreement his reasons for disagreement and his own valuation.

(3) If the Contractor does not give notice of disagreement or gives notice but fails to specify adequate grounds for disagreement he shall be deemed to have agreed the draft final account as the Final Account.

(4) If before the end of the Maintenance Period, or where there is more than one the end of the last Maintenance Period to expire, the Final Sum has been calculated and agreed or has been treated as having been agreed under paragraph (3), or in default of agreement has been determined in accordance with the Contract, then -

   (a) if the unpaid balance of the Final Sum exceeds any retention which the Employer is for the time being entitled to retain the Employer shall forthwith pay the excess to the Contractor; and

   (b) if the total amount paid to the Contractor exceeds the Final Sum the Contractor shall forthwith pay the excess to the Employer.

(5) If after the end of the Maintenance Period, or where there is more than one the end of the last Maintenance Period to expire, the PM has certified that the Contractor has complied with Condition 21 (Defects in Maintenance Periods), and the Final Sum has been calculated and agreed or has been treated as having been agreed under paragraph (3), or in default of agreement has been determined in accordance with the Contract, then -

   (a) if the Final Sum exceeds the amount previously paid to the Contractor, the Employer shall forthwith pay the excess to the Contractor; or

   (b) if the amount previously paid to the Contractor exceeds the Final Sum, the Contractor shall forthwith pay the excess to the Employer.

50

Certifying payments

(1) Subject to Conditions 59 (Adjudication) and 60 (Arbitration and choice of law), the PM shall certify in the prescribed form to the Employer, with a copy to the Contractor, all net sums (taking into account retention and all set-off or abatement to which the Employer is entitled, but exclusive of VAT) to which the Contractor and, if applicable, each nominated subcontractor and each nominated supplier, is entitled. The Contractor shall supply to the PM, not later than 7 Days before each such certificate is due, such information required to calculate the entitlements of each nominated subcontractor and each nominated supplier.

(2) Where Condition 48(2)(a) (Alternative A (Stage Payment Chart) or Alternative C (Valuation)) applies, the first such certificate shall be issued on a date to be agreed by the Employer and the Contractor, not later than 28 Days after the commencement of the execution of the Works, and subsequent certificates shall be issued on the equivalent date of each subsequent month, if any sum is to be certified.

(3) Where Condition 48(2)(a) (Alternative B (Milestone Payment Chart)) applies, each such certificate shall be issued within 21 Days of the PM's receipt of an application for payment under that Alternative, including (if the Abstract of Particulars so specifies) all amounts due under Condition 48(2)(b)-(h). If the Abstract of Particulars does not so specify, certification under paragraph (2) shall apply in respect of such amounts.

(4) Without prejudice to Condition 1(4)(c) (Definitions, etc.), if any date upon which a certificate should be issued is a Saturday or Sunday, or Christmas Day, Good Friday or a day which under the Banking and Financial Dealings Act 1971 is a bank holiday in England and Wales or, as the case may be, in Scotland or Northern Ireland, the relevant certificate shall be issued on the next working day.

(5) The Contractor shall immediately further copy each such certificate to each of his subcontractors and suppliers.

(6) After certification, the Contractor shall submit to the Employer a VAT invoice for the precise certified sum, plus applicable VAT. Subject to Condition 50A (Withholding payment), the Employer shall pay to the Contractor each certified sum, plus applicable VAT, within 30 Days of the Employer's receipt of an invoice, strictly in accordance with this Condition, in respect of that certified sum.

(7) Any certificate may be modified or corrected by any subsequent certificate or by the final certificate for payment. No certificate of the PM shall of itself be conclusive evidence that any work or Things to which it relates are in accordance with the Contract.

50A

Withholding payment

(1) The Employer or the Contractor, as the case may be, shall give notice not later than 5 Days after the date on which a payment becomes due from him under the Contract, or would have become due if -

   (a) the other party had carried out his obligations under the Contract; and

   (b) no set-off or abatement was permitted by reference to any sum claimed to be due under one or more other contracts;

specifying the amount (if any) of the payment made or proposed to be made, and the basis on which that amount was calculated.

(2) Neither the Employer nor the Contractor may withhold payment, after its due date for payment (which shall be its final date for payment for all purposes), of a sum due under the Contract, unless he has given an effective notice of intention to withhold payment. The notice mentioned in paragraph (1) may suffice as a notice of intention to withhold payment if it complies with the requirements of this paragraph. To be effective, a notice of intention to withhold payment must specify -

(a) the amount proposed to be withheld, and the ground for withholding payment; or

(b) if there is more than one ground, each ground and the amount attributable to it;

and must be given not later than 7 Days before the due date for payment of the relevant sum.

51
Recovery of sums

Without prejudice and in addition to any other rights and remedies of the Employer, whenever under or in respect of the Contract, or under or in respect of any other contract between the Contractor or any other member of the Contractor's Group and the Employer or any other member of the Employer's Group, any sum of money shall be recoverable from or payable by the Contractor or any other member of the Contractor's Group by or to the Employer or any other member of the Employer's Group, it may be deducted by the Employer from any sum or sums then due or which at any time thereafter may become due to the Contractor or any other member of the Contractor's Group under or in respect of the Contract, or under or in respect of any other contract between the Contractor or any other member of the Contractor's Group and the Employer or any other member of the Employer's Group. Without prejudice and in addition to any other rights and remedies of the Contractor, each member of the Contractor's Group shall have rights reciprocal to those of each member of the Employer's Group under this Condition.

PARTICULAR POWERS AND REMEDIES

52

Suspension for non-payment

(1) Where a sum due under the Contract (as determined by agreement between the parties, certification and invoicing, adjudication, arbitration or litigation) is not paid in full by the final date for payment and no effective notice to withhold payment has been given under Condition 50A (Withholding payment), the person to whom the sum is due has the right (without prejudice to any other right or remedy) to suspend performance of his obligations under the Contract to the party by whom payment ought to have been made ('the party in default').

(2) The right may not be exercised without first giving to the party in default at least 7 Days' notice of intention to suspend performance, stating the ground or grounds on which it is intended to suspend performance.

(3) The right to suspend performance ceases when the party in default makes payment in full of the amount due.

(4) Any period during which performance is suspended in pursuance of the right conferred by this Condition shall be disregarded in computing, for the purposes of any contractual time limit, the time taken, by the party exercising the right or by a third party, to complete any work directly or indirectly affected by the exercise of the right. Where the contractual time limit is set by reference to a date rather than a period, the date shall be adjusted accordingly.

53

Non-compliance with Instructions

If, after receipt of a notice from the PM requiring compliance with any Instruction within a period specified in the notice, the Contractor fails to comply, the Employer may, without prejudice to the exercise of his powers to determine the Contract, provide labour and/or any Things (whether or not for incorporation), or enter into a contract for the execution of any work which may be necessary to give effect to that Instruction. Any reasonable costs and expenses incurred by the Employer over and above those which would have been incurred had the Contractor complied promptly with the Instruction, shall be recoverable by the Employer from the Contractor.

54

Emergency work

(1) If the Contractor is unable or unwilling to carry out promptly any emergency work required by the PM, the Employer may make arrangements for that work to be carried out. If the work carried out by the Employer shall be such as the Contractor is liable under the Contract to carry out or execute at his own expense then the Contractor shall reimburse -

(a) any costs reasonably incurred by the Employer under this paragraph; and

(b) any loss suffered by the Employer because the Contractor has not carried out the work.

(2) In this Condition emergency work means any work which becomes necessary during the execution of the Works or during any Maintenance Period -

(a) to prevent, or alleviate the effects of, any accident, failure or other event in connection with the performance of the Works;

(b) to secure the Works, the Site or any adjoining property from damage; or

(c) without prejudice to Condition 19 (Loss or damage), to repair any damaged or dangerous part of the Works.

55

Liquidated damages

(1) This Condition applies where a rate of liquidated damages for any delay in the completion of the Works or a Section has been specified in the Abstract of Particulars.

(2) If the Works or a Section are or is not completed by the relevant Date for Completion, the Contractor shall immediately become liable to pay to the Employer liquidated damages at the rate specified in the Abstract of Particulars for the period that the Works or any relevant Section remain or remains uncompleted.

(3) Subject to Condition 50A (Withholding payment), the Employer may deduct any amount of liquidated damages to which he may be entitled under this Condition from any advances to which the Contractor may otherwise be entitled under Condition 48 (Advances on account).

(4) If the sum due as liquidated damages exceeds any advance payable to the Contractor under Condition 48 (Advances on account), the Contractor shall pay to the Employer the difference. That sum shall be recoverable in accordance with Condition 51 (Recovery of sums).

(5) No payment or concession to the Contractor, or Instruction or VI at any time given to the Contractor (whether before or after the Date or Dates for Completion), or other act or omission by or on behalf of the Employer, shall in any way affect the rights of the Employer to deduct or recover liquidated damages, or shall be deemed to be a waiver of the right of the Employer to recover such damages. The rights of the Employer to deduct or recover liquidated damages may be waived only by notice from the Employer to the Contractor.

56

Determination by Employer

(1) Without prejudice to any other power of determination, the Employer may determine the Contract by notice to the Contractor if -

(a) any ground mentioned in subparagraphs (6) (a), (b) or (e) has arisen; the Employer has given notice to the Contractor specifying the relevant ground and facts; and such ground was in existence 14 Days after such notice was given; or has arisen again at any subsequent time; or

(b) any ground mentioned in paragraph (6) has arisen, other than those mentioned in subparagraphs (6) (a), (b) and (e). (2) The Employer shall specify in a notice of determination under paragraph (1) which of the grounds mentioned in paragraph (6) apply.

(3) The Employer may, after giving notice of determination under paragraphs (1) or (8), give directions in relation to the performance or completion of any work and any other matters connected with the Works, the Site and any other contract or subcontract.

(4) Any directions under paragraph (3) shall be given not later than three months from the date of the notice of determination under paragraphs (1) or (8) or the relevant Date for Completion, whichever is the sooner.

(5)   The Contractor shall comply promptly with any directions given by the Employer under paragraph (3). The Contractor shall be paid for any work so performed as if the directions were Instructions.

(6)   The grounds referred to in paragraph (1) are -

(a)   the failure of the Contractor to comply with an Instruction within a reasonable period of its issue;

(b)   the failure of the Contractor to execute work in a workmanlike or proper manner, or to proceed regularly and diligently with the Works, or the suspension by the Contractor of the Works (otherwise than in accordance with Condition 52 (Suspension for non-payment)), so that in the opinion of the PM the Contractor has not completed or will be unable to secure the completion of the Works or any relevant Section by the Dates for Completion;

(c)   where the Contractor is an individual, the insolvency of that individual or, where the Contractor is a partnership, the insolvency of any partner; in this subparagraph 'insolvency' shall include the presentation of a bankruptcy petition where the petitioner is the debtor; the making of a bankruptcy order; the appointment of an interim receiver; the issue of proposals by the debtor to creditors for any arrangement or composition with creditors (whether as a voluntary arrangement under Part VIII of the Insolvency Act 1986 or of the Insolvency (Northern Ireland) Order 1989, or otherwise) or for a conveyance or assignment for the benefit of creditors (whether under the Deeds of Arrangement Act 1914 or otherwise); where the Contractor is a partnership, insolvency shall also include the presentation of a petition by the members of the partnership to wind up the partnership as an unregistered company under Part V of the Insolvency Act 1986 or Part VI of the Insolvency (Northern Ireland) Order 1989; the making of an order to wind up the partnership as an unregistered company; the presentation by the members of the partnership of a petition for the making of an administration order in respect of the partnership; the making of an administration order; or the issue of proposals by the members of the partnership to its creditors for a voluntary arrangement; in Northern Ireland, the term insolvency shall also include in respect of the Contractor, if an individual, or the partnership or any partner if the Contractor is a partnership, becoming insolvent within the meaning of Article 12 of the Construction Contracts (Northern Ireland) Order 1997 (whether or not in force); and in Scotland, the term insolvency shall also include in respect of the Contractor, if an individual, or the partnership or any partner if the Contractor is a partnership, becoming bankrupt, or having his or its estate sequestrated, or becoming apparently insolvent as defined in the Bankruptcy (Scotland) Act 1985, or entering into a trust deed for his or its creditors, or making a composition or arrangement with his or its creditors;

(d)   where the Contractor is a company then (in respect of that company or any company which is for the time being a holding company of the Contractor) the presentation by the company or its directors of a petition for winding-up, or the passing of any resolution for the winding-up, of the relevant company (except for the purposes of amalgamation or reconstruction while solvent); the making by the court of a winding-up order; the appointment of a provisional liquidator; the presentation by the company or its directors of a petition for or the making of an administration order, or a railway administration order; the issue by the company of proposals to creditors for the making of any arrangement or composition with creditors (whether as a voluntary arrangement under Part I of the Insolvency Act 1986, or Part II of the Insolvency (Northern Ireland) Order 1989, or a scheme of arrangement under Section 425 of the Companies Act 1985, or Article 418 of the Companies (Northern Ireland) Order 1986, or otherwise); the appointment of an administrative receiver or a receiver or a receiver and manager in respect of the company or of any of its assets; or the relevant company becoming insolvent within the meaning of Article 12 of the Construction Contracts (Northern Ireland) Order 1997 (whether or not in force);

(e) failure by the Contractor to comply with Condition 26 (Site admittance) where the Employer determines that such failure is prejudicial to the interests of the Employer or any other member of the Employer's Group;

(f) either of the situations described in Condition 24(3)(a) or (b) (Corruption);

(g) failure by the Contractor to comply with Condition 66 (Performance bond) (if applicable) or Condition 67 (Parent company guarantee) (if applicable), in respect of each of which Conditions time shall be of the essence; or

(h) any breach of the conditions set out in the invitation to tender relating to the Contract.

(7) All Things not for incorporation which are brought onto the Site at the Contractor's expense shall (whether damaged or not) re-vest in and be removed by him as and when they cease to be required in connection with any directions given by the Employer under paragraph (3). From the date of determination the Employer shall be under no liability to the Contractor in respect of any loss or damage to any such Things caused by any of the Accepted Risks.

(8) Without prejudice to any other power of determination, the Employer may at will determine the Contract by notice to the Contractor. If the Employer purports to determine the Contract under paragraph (1), but no ground for determination under paragraph (1) had then arisen, or any such ground which had arisen had been waived by the Employer, the Employer shall be deemed to have decided to determine, and to have determined, the Contract under this paragraph.

57

Consequences of determination by Employer

(1) If the Employer shall determine the Contract for any reason mentioned in Condition 56(6) (Determination by Employer), the following provisions shall apply -

(a) all sums of money that may then be due or accruing due from the Employer to the Contractor shall cease to be due or to accrue due;

(b) the Employer may hire any person, employ other contractors, use any Things on the Site, and may purchase or do anything necessary for the completion of the Works, and the Contractor shall have no claim whatsoever in respect of any such action by the Employer;

(c) the Contractor shall (except where determination occurs by reason of any of the circumstances described in Condition 56(6)(c) and (d) (Determination by Employer)) assign to the Employer, without further payment, the benefit of any subcontract or contract for the supply of any Thing for incorporation which he may have made in connection with the Contract;

(d) the Employer may pay to any subcontractor or supplier any amount due to him which the PM certifies as included in any previous advance to the Contractor, and the amount so paid shall be forthwith recoverable by the Employer from the Contractor; and

(e) the QS shall ascertain and the PM shall certify the cost to the Employer of completion of the Works.

(2)   If the total of the following sums (hereafter called 'the first amount') exceeds the total of all the advances paid to the Contractor (or to which he is entitled) under Condition 48 (Advances on account) to the date of determination (hereafter called the second amount), the Employer shall, subject to paragraph (3), hold the amount of the excess. If the second amount exceeds the first amount the Contractor shall be liable to pay the Employer the amount of the excess. The individual sums are -

(a)   the value of all the work carried out in accordance with the Contract up to the date of determination;

(b)   the value of any work carried out or other things done in accordance with any direction given under Condition 56(3) (Determination by Employer); and

(c)   the value (ascertained on the basis of fair and reasonable prices) of all Things for incorporation brought onto the Site, or in the course of preparation or manufacture off the Site, which the Employer elects to keep.

(3)   If the total of the cost of completion as certified under subparagraph (1)(e) and the first amount determined in accordance with paragraph (2) exceeds the sum that would have been payable to the Contractor for due completion of the Works, then the Contractor shall pay the Employer the amount of the excess. If the total of the cost of completion as certified under subparagraph (1)(e) and the first amount determined in accordance with paragraph (2) is less than the sum that would have been payable to the Contractor for due completion of the Works, then the Employer shall pay the Contractor the amount of the shortfall.

(4)   If the Employer shall determine, or shall be deemed to have determined, the Contract at will under Condition 56(8) (Determination by Employer), Condition 58(5) and (6) (Determination by Contractor) shall apply, as if the Contractor had determined the Contract under that Condition.

58

Determination by Contractor

(1)   Without prejudice to any other power of determination, the Contractor may determine the Contract by notice to the Employer if -

(a)   the ground mentioned in subparagraph (3)(a) has arisen; the Contractor has consequently suspended performance of his obligations under the Contract under Condition 52 (Suspension for non-payment) for a continuous period of not less than 30 Days; following which period the Contractor has given notice to the Employer specifying the relevant ground and facts; such ground was in existence 14 Days after such notice was given; and the above suspension is still continuing, and has continued without a break since its commencement; or

(b)   any ground mentioned in subparagraphs (3)(b) or (e) has arisen; the Contractor has given notice to the Employer specifying the relevant ground and facts; and such ground was in existence 14 Days after such notice was given; or has arisen again at any subsequent time; or

(c)   any ground mentioned in paragraph (3)(c) or (d) has arisen;

(2)   The Contractor shall specify in a notice of determination which of the grounds mentioned in paragraph (3) apply.

(3)   The grounds referred to in paragraph (1) are -

(a)   the failure of the Employer to pay any sum certified by the PM for payment to the Contractor (and an appropriate sum in respect of VAT under Condition 45 (VAT)) within the time allowed under Condition 50 (Certifying payments);

(b) the Employer obstructing, or interfering with, the issue of any certificate of the PM referred to in Condition 50(1) (Certifying payments);

(c) where the Employer is an individual, the insolvency of that individual or, where the Employer is a partnership, the insolvency of any partner; in this subparagraph insolvency shall include the occurrence in respect of the individual or partner of any of the matters referred to in Condition 56(6)(c) (Determination by Employer);

(d) where the Employer is a company, the occurrence in respect of that company or any company which is for the time being a holding company of the Employer of any of the matters referred to in Condition 56(6)(d) (Determination by Employer); or

(e) the suspension of the execution of the whole or substantially the whole of the Works (other than the making good of defects in the Works under Condition 21 (Defects in Maintenance Periods) for a continuous period of 182 Days (or such other period as is stated in the Abstract of Particulars)) by reason of -

   (i) any Instruction, except where the Instruction was necessary because of any default or neglect by the Contractor or by any employee, agent or subcontractor of his; or

   (ii) any of the matters referred to in Condition 46(1)(a), (b) or (c) (Prolongation and disruption).

(4) If the Contractor shall determine the Contract as mentioned in this Condition, or if the Employer shall determine, or shall be deemed to have determined, the Contract at will under Condition 56(8) (Determination by Employer) and Condition 57(4) (Consequences of determination by Employer) consequently applies, the following provisions shall apply.

(5) If the total of the following sums (hereafter called 'the third amount') exceeds the second amount determined in accordance with Condition 57(2) (Consequences of determination by Employer), the Employer shall be liable to pay the Contractor the amount of the excess. If the second amount exceeds the third amount the Contractor shall be liable to pay the Employer the amount of the excess. The individual sums are -

(a) the value of all the work carried out in accordance with the Contract up to the date of determination;

(b) the value of any work carried out or other things done in accordance with any direction given under Condition 56(3) (Determination by Employer);

(c) the value (ascertained on the basis of fair and reasonable prices) of all Things for incorporation brought onto the Site, or in the course of preparation or manufacture off the Site, which the Employer elects to keep;

(d) any reasonable sum expended by the Contractor because of the determination of the Contract in respect of -

   (i) the uncompleted part of any subcontract and other contracts (including those for the hire of plant, services and insurance); and

> (ii) any unavoidable contract of employment, entered into in connection with the Contract; and

> (e) the Contractor's other unavoidable losses or expense directly due to the determination (including, without limitation, loss of profit on the Contract).

(6) All Things not for incorporation which are brought onto the Site at the Contractor's expense shall (whether damaged or not) re-vest in and be removed by him. From the date of determination the Employer shall be under no liability to the Contractor in respect of any loss or damage to any such Things caused by any of the Accepted Risks.

58A

Determination following suspension of Works

(1) Without prejudice to any other power of determination, either the Employer or the Contractor may determine the Contract by notice to the other of them in the event of the suspension of the execution of the whole or substantially the whole of the Works (other than the making good of defects in the Works under Condition 21 (Defects in Maintenance Periods)) for a continuous period of 182 Days (or such other period as is stated in the Abstract of Particulars): provided that such suspension for that period was by reason of any circumstances, not arising because of any act, neglect or default of the Contractor or of any employee, agent or subcontractor of his; and not arising because of any act, neglect or default of the Employer or the PM or of any other contractor or agent of the Employer; and other than weather conditions; which are outside the control of the Employer and of the Contractor or any of his subcontractors and which could not have been reasonably contemplated under the Contract (including, without limitation, loss or damage to the Works or any Things on the Site caused by an Accepted Risk).

(2) The party giving notice of determination under paragraph (1) shall specify in the notice the grounds for such determination.

(3) If either the Employer or the Contractor shall determine the Contract as mentioned in this Condition, then Condition 58(5) and (6) (Determination by Contractor) shall apply, provided that the Contractor shall not be entitled to any loss of profit on the Contract.

59

Adjudication

(1) The Employer or the Contractor may at any time notify the other of intention to refer a dispute, difference or question arising under, out of, or relating to, the Contract to adjudication. Within 7 Days of such notice, the dispute, may by further notice be referred to the adjudicator specified in the Abstract of Particulars.

(2) The notice of referral shall set out the principal facts and arguments relating to the dispute. Copies of all relevant documents in the possession of the party giving the notice of referral shall be enclosed with the notice. A copy of the notice and enclosures shall at the same time be sent by the party giving the notice to the PM, the QS and the other party.

(3) (a) If the person named as adjudicator in the Abstract of Particulars is unable to act, or is not or ceases to be independent of the Employer, the Contractor, the PM and the QS, he shall be substituted as provided in the Abstract of Particulars.

> (b) It shall be a condition precedent to the appointment of an adjudicator that he shall notify both parties that he will comply with this Condition and its time limits.

(c) The adjudicator, unless already appointed, shall be appointed within 7 Days of the giving of a notice of intention to refer a dispute to adjudication under paragraph (1). The Employer and the Contractor shall jointly proceed to use all reasonable endeavours to complete the appointment of the adjudicator and named substitute adjudicator. If either or both such joint appointments has not been completed within 28 Days of the acceptance of the tender, either the Employer or the Contractor alone may proceed to complete such appointments. If it becomes necessary to substitute as adjudicator a person not named as adjudicator or substitute adjudicator in the Abstract of Particulars, the Employer and Contractor shall jointly proceed to use all reasonable endeavours to appoint the substitute adjudicator. If such joint appointment has not been made within 28 Days of the selection of the substitute adjudicator, either the Employer or Contractor alone may proceed to make such appointment. For all such appointments, the form of adjudicator's appointment prescribed by the Contract shall be used, so far as is reasonably practicable. A copy of each such appointment shall be supplied too each party. No such appointment shall be amended or replaced without the consent of both parties.

(4) The PM, the QS and the other party may submit representations to the adjudicator not later than 7 Days from the receipt of the notice of referral.

(5) The adjudicator shall notify his decision to the PM, the QS, the Employer and the Contractor not earlier than 10 and not later than 28 Days from receipt of the notice of referral, or such longer period as is agreed by the Employer and the Contractor after the dispute has been referred. The adjudicator may extend the period of 28 Days by up to 14 Days, with the consent of the party by whom the dispute was referred. The adjudicator's decision shall nevertheless be valid if issued after the time allowed. The adjudicator's decision shall state how the cost of the adjudicator's fee or salary (including overheads) shall be apportioned between the parties, and whether one party is to bear the whole or part of the reasonable legal and other costs and expenses of the other, relating to the adjudication.

(6) The adjudicator may take the initiative in ascertaining the facts and the law, and the Employer and the Contractor shall enable him to do so. In coming to a decision the adjudicator shall have regard to how far the parties have complied with any procedures in the Contract relevant to the matter in dispute and to what extent each of them has acted promptly, reasonably and in good faith. The adjudicator shall act independently and impartially, as an expert adjudicator and not as an arbitrator. The adjudicator shall have all the powers of an arbitrator acting in accordance with Condition 60 (Arbitration and choice of law), and the fullest possible powers to assess and award damages and legal and other costs and expenses; and, in addition to, and notwithstanding the terms of, Condition 47 (Finance charges), to award interest. In particular, without limitation, the adjudicator may award simple or compound interest from such dates, at such rates and with such rests as he considers meet the justice of the case -

(a) on the whole or part of any amount awarded by him, in respect of any period up to the date of the award;

(b) on the whole or part of any amount claimed in the adjudication proceedings and outstanding at the commencement of the adjudication proceedings but paid before the award was made, in respect of any period up to the date of payment;

and may award such interest from the date of the award (or any later date) until payment, on the outstanding amount of any award (including any award of interest and any award of damages and legal and other costs and expenses).

(7) Subject to the proviso to Condition 60(1) (Arbitration and choice of law), the decision of the

adjudicator is binding until the dispute is finally determined by legal proceedings, by arbitration (if the Contract provides for arbitration, or the parties otherwise agree to arbitration), or by agreement: and the parties do not agree to accept the decision of the adjudicator as finally determining the dispute.

(8) In addition to his other powers, the adjudicator shall have power to vary or overrule any decision previously made under the Contract by the Employer, the PM or the QS, other than decisions in respect of the following matters -

    (a) decisions by or on behalf of the Employer under Condition 26 (Site admittance);

    (b) decisions by or on behalf of the Employer under Condition 27 (Passes) (if applicable);

    (c) provided that the circumstances mentioned in Condition 56(1)(a) or (b) (Determination by Employer) have arisen, and have not been waived by the Employer, decisions of the Employer to give notice under Condition 56(1)(a), or to give notice of determination under Condition 56(1);

    (d) decisions or deemed decisions of the Employer to determine the Contract under Condition 56(8) (Determination by Employer);

    (e) provided that the circumstances mentioned in Condition 58A(1) (Determination following suspension of Works) have arisen, and have not been waived by the Employer, decisions of the Employer to give notice of determination under Condition 58A(1); and

    (f) decisions of the Employer under Condition 61 (Assignment).

In relation to decisions in respect of those matters, the Contractors's only remedy against the Employer shall be financial compensation.

(9) Notwithstanding Condition 60 (Arbitration and choice of law), the Employer and the Contractor shall comply forthwith with any decision of the adjudicator; and shall submit to summary judgment and enforcement in respect of all such decisions.

(10) If requested by one of the parties to the dispute, the adjudicator shall provide reasons for his decision. Such requests shall only be made within 14 Days of the decision being notified to the requesting party.

(11) The adjudicator is not liable for anything done or omitted in the discharge or purported discharge of his functions as adjudicator, unless the act or omission is in bad faith. Any employee or agent of the adjudicator is similarly protected from liability.

60

Arbitration and choice of law

(1) The procedures for arbitration set out in this Condition shall be utilised by the Employer and the Contractor with regard to disputes, differences or questions between the Employer and the Contractor arising out of, or relating to, the Contract, whether during the course or after the determination thereof (other than disputes, differences or questions arising out of, or relating to, the enforceability or enforcement of any adjudicator's decision; or Condition 45 (VAT); or the statutory tax deduction scheme referred to in the Abstract of Particulars). The dispute, difference or question shall after notice by either party to the other be referred to the single arbitrator specified in the Abstract of Particulars. If the person named in the Abstract of Particulars is deceased or unwilling or unable to act as arbitrator, or is not or ceases to be independent of the

Employer, the Contractor, the PM and the QS, he shall be substituted as provided in the Abstract of Particulars. It shall be a condition precedent to the appointment of an arbitrator that he shall notify both parties that he will forthwith commence his duties, and will comply with this Condition and its time limits. In addition to his other powers, the arbitrator shall have the fullest possible powers -

(a) to rectify the Contract;

(b) to order inspections, measurements and valuations;

(c) to vary or overrule any decision previously made under the Contract by the Employer, the PM, the QS or an adjudicator, provided that the Contractor's only remedy against the Employer in relation to decisions in respect of the matters listed in Condition 59(8)(a)-(f) (Adjudication) shall be financial compensation;

(d) to order consolidation of the proceedings with other proceedings; and/or that concurrent hearings shall be held; and to make such orders and directions relating to such consolidation and hearings as he thinks fit; and

(e) to make orders, directions and awards in the same way as if all the procedures of the High Court of Justice in England (if the proper law of the Contract is English law), or the High Court of Justice in Northern Ireland (if the proper law of the Contract is Northern Ireland law), or the Court of Session in Scotland (if the proper law of the Contract is Scots law), including, without limitation, as to joining one or more defendants or defenders or joining co- defendants or co-defenders or third parties, were available to the parties and to the arbitrator.

Provided always, that where any dispute, difference or question has been referred to an adjudicator under Condition 59 (Adjudication), and the adjudicator has issued his decision thereon, a party shall not be entitled to refer such dispute, difference or question to arbitration, and the adjudicator's decision thereon shall become unchallengeable, unless that party serves the above notice within 56 Days of receipt of notification of the adjudicator's decision: and, for the avoidance of doubt, this proviso shall apply whether or not the adjudicator has notified his decision within the time limit specified in Condition 59(5).

(2) Unless the parties otherwise agree -

(a) no reference shall be made under paragraph (1) until after the completion, alleged completion or abandonment of the Works, or the determination of the Contract;

(b) no reference shall be made under paragraph (1) in respect of a dispute, difference or question which has been referred to an adjudicator under Condition 59 (Adjudication), until the notification by the adjudicator of his decision thereon; or the expiry of 28 Days from the receipt by the adjudicator of the notice of referral under Condition 59(1), or such longer time as is allowed for the adjudicator's decision in accordance with Condition 59; whichever is the earlier;

(c) the arbitrator shall hold a preliminary meeting with the parties forthwith on his acceptance of office and will fix a timetable for the delivery of points of claim, defence or counter-claim and any other pleadings, for the discovery and inspection of documents, for the inspection of the Works (if necessary) and for the hearing of any oral evidence (if necessary), which timetable shall not without the consent of the parties exceed a period of 6 months from the date of that preliminary meeting;

(d) the parties shall ensure that any evidence whether oral or written and any document or argument required to be submitted to the arbitrator is submitted to him in accordance with the timetable; and

(e) the arbitrator shall give his award not later than 3 months from the end of the period mentioned in subparagraph (c).

(3) (a) If the Works are in England or Wales, the proper law of the Contract shall be English law, a reference to arbitration in accordance with this Condition shall be a reference to which the Arbitration Act 1996 applies, and that Act shall have effect subject to the provisions of this Condition.

(b) If the Works are in Scotland, the Contract shall in all respects be construed and operated as a Scottish contract, and shall be interpreted in accordance with Scots law.

(c) If the Works are in Northern Ireland, the proper law of the Contract shall be Northern Ireland law, a reference to arbitration in accordance with this Condition shall be a reference to which the Arbitration Act 1996 applies, and that Act shall have effect subject to the provisions of this Condition. Whatever the nationality, residence or domicile of the Contractor, any sub-contractor or supplier, or the arbitrator, such arbitration shall be conducted solely within Northern Ireland.

(4) If the Contract is subject to Scots law, the following shall apply-

(a) References throughout the Contract to a 'Deed' or 'Deeds' shall be construed as references to a document or documents subscribed in accordance with the requirements for a self-proving document under the Requirements of Writing (Scotland) Act 1995.

(b) Condition 48C (Payment for Things off-Site) shall not apply.

(c) In Condition 59(9) (Adjudication), the words 'submit to summary judgement and' shall be substituted by the words 'consent to a motion for summary decree and submit to': and there shall also be added at the end the following sentence:

'Where the Employer, the Contractor or the adjudicator wishes to register the decision of the adjudicator for execution in the Books of Council and Session, any other party shall, on being requested to do so, forthwith consent to such registration by subscribing the decision before a witness'.

(d) In subparagraph (2)(c), the words 'points of claim' shall be substituted by the words the 'statement of claim'; and the word 'discovery' shall be substituted by the word 'recovery'.

(e) The law of Scotland shall apply to any adjudication and to any arbitration, and any arbitration shall be conducted in accordance with the Scottish Arbitration Rules of the Chartered Institute of Arbitrators (Arbiters) (Scottish Branch) 1996 Edition, or such amended Scottish Rules as the said Chartered Institute may have adopted to take effect before the commencement of the arbitration, the terms of which are deemed to be incorporated herein, provided that to the extent that there is inconsistency between the terms of those Rules and this Condition, the latter shall have precedence.

(f) Subject to the provisions of the Contract in respect of adjudication and arbitration, the parties submit to the non-exclusive jurisdiction of the Scottish courts.

ASSIGNMENT, SUBLETTING, SUBCONTRACTING, SUPPLIERS AND OTHER WORKS

61

Assignment

The Contractor shall not, without the consent in writing of the Employer, assign or transfer the Contract, or any part, share or interest under it. No sum of money to become payable under the Contract shall be payable to any person other than the Contractor without the Employer's written consent. The Employer may assign or transfer the benefit of the Contract, or any part, share or interest under it. If so provided by the Abstract of Particulars, this power shall only be exercisable after certification by the PM under Condition 39 (Certifying completion) of the completion of the Works or the last Section thereof in respect of which completion is certified, or the determination of the Contract for any reason whatsoever, including (without limitation) breach by the Employer, whichever is the earlier. Neither party may assign or transfer the whole or any part of the burden of the Contract.

62

Subletting

(1) (a) Except where the Employer has accepted a subletting proposal prior to the award of the Contract, or the Contract specifies or nominates the subletting of work, the Contractor shall not sublet any part of the Contract without the prior consent of the PM.

(b) The Contractor shall provide such details of any subcontractor he wishes to engage, and (other than rates and prices) of any subcontract entered into, as the PM may require.

(2) The Contractor shall ensure that each subcontract entered into will enable him to fulfil his obligations under the Contract. No subcontract shall include any provision making payment under the subcontract conditional on the payer receiving payment from a third person unless that third person, or any other person payment by whom is under the subcontract (directly or indirectly) a condition of payment by that third person, is insolvent within the meaning of Section 113 of the Housing Grants, Construction and Regeneration Act 1996, or, if the Works are in Northern Ireland, of Article 12 of the Construction Contracts (Northern Ireland) Order 1997 (whether or not in force). Each subcontract shall include -

(a) Power to determine the subcontract as a consequence of determination of this Contract under Conditions 56 (Determination by Employer), 57 (Consequences of determination by Employer), 58 (Determination by Contractor) or 58A (Determination following suspension of Works).

(b) A provision to the effect that from the commencement to the completion of the subcontract work all Things belonging to the person who enters into the subcontract which are brought on the Site in connection with the subcontract shall vest in the Contractor subject to any right of the Contractor to reject the same.

(c) Such provisions as may be necessary to enable the Contractor to fulfil his obligations to the Employer under the Contract.

(d) Such provisions as will impose on the person who enters into the subcontract liabilities similar to those imposed on the Contractor under the Contract.

(e) A provision to the effect that no part of the subcontract work shall be further sublet without the consent of the Contractor.

(f) Provisions similar to Condition 38(4)-(7) (Acceleration and cost savings), so that the Contractor shall share fairly and equitably with the subcontractor the Contractor's part of savings pursuant thereto; addressing the sharing of such savings between the Contractor and the subcontractor by reference to whether the relevant proposal for cost savings originated with the Contractor or the subcontractor.

(g) Provisions equivalent to Conditions 1A (Fair dealing and teamworking), 50A (Withholding payment), 52 (Suspension for non-payment) and 59 (Adjudication).

(h) Provisions entitling the subcontractor to payment by instalments, stage payments or other periodic payments for any work under the subcontract unless it is specified in the subcontract that the duration of the work is to be less than 45 Days; or it is agreed between the Contractor and the subcontractor that the duration of the work is estimated to be less than 45 Days; and stipulating the amounts of the payments and the intervals at which, or circumstances in which, they become due.

(i) Terms and conditions providing an adequate mechanism for determining what payments become due under the subcontract, and when; providing for a final date for payment in relation to any sum which becomes due; and stipulating how long the period (not longer than 30 Days) is to be between the date on which a sum becomes due and the final date for payment.

(j) Any subcontract terms and conditions, or form of subcontract, prescribed in the Abstract of Particulars.

(3) Without prejudice to the obligations of the Contractor under any of the provisions of the Contract, the Contractor shall, whenever requested to do so by the Employer, take any necessary action to ensure that a person who has entered into a subcontract complies with and performs all obligations imposed upon him.

(4) Where for any reason a subcontract is determined or assigned because of the default or failure of the subcontractor, the Contractor shall, subject to Condition 63A (Insolvency of nominated subcontractors or suppliers) (if applicable), at his own expense secure completion of the sub-contract work.

(5) The Contractor shall be responsible for any subcontractor or supplier employed by him in connection with the Works, whether or not nominated or approved by the Employer or the PM, or appointed by the Contractor in accordance with an Instruction or otherwise.

(6) Subject to Condition 63A (Insolvency of nominated subcontractors or suppliers) (if applicable) the Contractor shall make good any loss suffered or expense reasonably incurred by the Employer by reason of any default or failure, whether total or partial, on the part of any subcontractor or supplier.

63

Nomination

(1) A nominated subcontractor or nominated supplier means a person with whom the Contractor is required to enter into a contract for the execution of work or the supply of Things designated as Prime Cost items by the use of the expression PC or otherwise. This requirement may be specified in the Contract documents or in any direction or Instruction given under the Contract.

(2) All Prime Cost items designated as above shall be reserved for execution or supply by a person to be nominated or appointed in such ways as may be directed by the Employer or Instructed by the PM. The Contractor shall not without the written consent of the PM order work or Things under such items prior to the conclusion of an authorised subcontract. The Employer reserves the right to order and pay for all or any part of such items direct, and to deduct these payments from the Contract Sum less an amount in respect of Contractor's profit at the rate included in the Schedule of Rates adjusted pro-rata on the amount paid direct by the Employer.

(3) The sum to be paid by the Employer in respect of any Prime Cost or PC item shall be the sum (inclusive of charges for packing, carriage and delivery to the Site and after the deduction of all discounts, rebates or allowances) properly due to the nominated subcontractor or supplier after adjustment in respect of overpayment or overmeasurement. Any resulting increase or decrease in the Prime Cost sum shall be added to or deducted from the Contract Sum. The Contractor shall produce to the QS such quotations, invoices and bills (properly receipted) as may be necessary to show details of the actual sums paid by the Contractor.

(4) In addition to payments under paragraph (3), the Contractor shall be entitled to payment for fixing any Thing supplied by a nominated supplier in accordance with the rates included in the Schedule of Rates, and to profit. The payment for fixing shall cover unloading, getting-in, unpacking, return of empties and other incidental expenses. The Contractor's profit shall be adjusted pro-rata on the Prime Cost, excluding any alterations in that Prime Cost item due to the operation of any conditions incorporated in the nominated subcontract pursuant to Condition 62(2) (Subletting).

(5) If any Work or Things to which a nominated subcontract applies is not included in the Specification the Contractor shall, if required by the Employer, supply a full and detailed schedule of rates, which was properly and reasonably used for calculating the Contract Sum or subcontract sum, such schedule to be used in place of the Schedule of Rates for valuation and measurement under the Conditions of Contract.

(6) The Employer shall not require the Contractor to enter into a subcontract with any subcontractor against whom the Contractor has made a reasonable objection, except that the Contractor shall not object to any subcontractor or supplier nominated in the documents provided by the Employer as the basis of the Contractor's tender. The Contractor shall provide such information as the PM may reasonably require in relation to any objection which he makes under this paragraph.

(7) Once the Contractor has entered into a subcontract with a nominated subcontractor or supplier, he shall not determine or assign that subcontract without the agreement of the Employer. Where the Employer has agreed to the determination of the subcontract, the Employer shall as soon as reasonably practicable either nominate a replacement subcontractor or supplier, or direct the Contractor to complete the work or supply in question with his own resources or by a subcontractor or supplier of his own choice approved by the Employer.

(8) Subject to Condition 63A (Insolvency of nominated subcontractors or suppliers) (if applicable), if a nominated subcontract is determined or assigned or re- nomination occurs, the Employer shall not be required to pay the Contractor any greater sums than would have been payable if such determination, assignment or re-nomination had not occurred.

Insolvency of nominated subcontractors or suppliers (only applicable if stated in Abstract of Particulars)

If a subcontract is determined or assigned because the nominated subcontractor or supplier becomes insolvent (and insolvency for this purpose shall include the matters referred to in Condition 56(6)(c) and (d) (Determination by Employer) as if references therein to the Contractor were references to the nominated subcontractor or supplier), the Employer shall reimburse the Contractor an amount equal to the difference between -

(a) any cost he has incurred in securing the completion of the subcontract work which exceeds what the cost to him of completing such subcontract work would have been under the original subcontract; and

(b) the amount which by using his best endeavours he has or should have recovered from the original subcontractor.

64

Provisional sums

The full amount of any provisional lump sums included in the Contract and the net value annexed to each of the provisional items inserted in the Schedule of Rates shall be deducted from the Contract Sum, and the value of work ordered and executed thereunder shall be ascertained as provided by Condition 42 (Valuation of Variation Instructions). No work under these items is to be commenced without Instruction from the PM.

65

Other works

(1) The Employer shall have power at any time to execute other works (whether or not in connection with the Works) on the Site at the same time as the Works are being executed. The Contractor shall give reasonable facilities for these works.

(2) The Contractor shall not be responsible for damage done to other works except for damage caused by the negligence, omission or default of his workpeople, agents or subcontractors. Any damage done to the Works in the execution of other works shall, for the purposes of Condition 19 (Loss or damage), be deemed to be damage which is wholly caused by the neglect or default of the Employer or of any other contractor or agent of the Employer.

PERFORMANCE BOND, PARENT COMPANY GUARANTEE AND COLLATERAL WARRANTIES

66

Performance bond (only applicable if stated in Abstract of Particulars)

The Contractor shall, within 21 Days of the acceptance of the tender, deliver to the Employer a performance bond in the form prescribed by the Contract from the surety or sureties named in the tender, in an amount of 10% of the Contract Sum (or such other percentage as is stated in the Abstract of Particulars).

67

Parent company guarantee (only applicable if stated in Abstract of Particulars)

The Contractor shall, within 21 Days of the acceptance of the tender, deliver to the Employer a parent company guarantee in the form prescribed by the Contract from its ultimate holding company (if any) named in the tender.

68

Collateral warranties (only applicable if stated in Abstract of Particulars)

(1) (a) The Contractor shall use reasonable endeavours to procure that each subcontract or supply contract with each nominated subcontractor or supplier, and (unless waived by the Employer) any other subcontractor or supplier, shall contain obligations on the relevant subcontractor or supplier to execute, in favour of the Employer, a Deed in the form prescribed by the Contract, or a similar form reasonably required by the Employer, and deliver the same to the Employer; together in each case (unless waived by the Employer) with a guarantee in the form prescribed by the Contract, or a similar form reasonably required by the Employer, from the ultimate holding company (if any) of the relevant subcontractor or supplier in respect of the subcontractor's or supplier's obligations pursuant to such Deed. The Contractor shall enforce such obligations, or such modified obligations as are referred to below.

(b) If, despite the Contractor having used such reasonable endeavours, the subcontractor or supplier will not accept such obligations, or will only accept them in a modified form, the Contractor shall notify the Employer, who may agree in writing that the relevant subcontract or supply contract need not contain such obligations, or that the relevant obligations may be in a modified form agreeable to the subcontractor or supplier.

(c) Failing such agreement by the Employer in the case of a proposed nominated subcontractor or supplier, the Contractor shall not be obliged to enter into a subcontract or supply contract with that nominated subcontractor or supplier.

(d) Failing such agreement by the Employer in the case of any other subcontractor or supplier, the Contractor shall not enter into a relevant subcontract or supply contract with that subcontractor or supplier.

(2) The above obligations for the provision of Deeds and guarantees shall continue notwithstanding determination of the Contract for any reason whatsoever, including (without limitation) breach by the Employer.

SCHEDULE OF TIME LIMITS

Condition	Subject	Limit	Imposed on
8(4)	Send other party an insurance certificate	21 days from tender acceptance or renewal/expiry date	Insuring Party
21(4)	Defects which have been made good	Relevant Maintenance Period	Contractor
35(2)	Hold progress meetings	Monthly	PM
35(3)	Submit report to PM prior to each progress meeting	5 days	Contractor
35(4)	Issue a statement of progress after each progress meeting	7 days	PM
36(1)	Notify Contractor of decisions regarding extension of time for whole or part of Works	42 days from date of notice by Contractor	PM
36(4)	Make final decision on all outstanding and interim extensions of time	42 days after completion of Works	PM
36(5)	Notify PM of dissatisfaction with decision	14 days from receipt of decision	Contractor
36(5)	Notify Contractor of any amendment to decision	28 days from receipt of notification	PM
40(3)	Confirm oral Instructions in writing	7 days	PM
40(5)	Submit to the QS a lump sum quotation showing the cost of complying with a VI	21 days from receipt of VI	Contractor
42(3)	Decide on lump sum quotation	21 days from receipt	PM
42(7)	Submit information required by Condition 41(4)	14 days from request by QS	Contractor
42(8)	Notify Contractor of valuation of a VI	28 days from receipt of information under Condition 42(7)	QS
42(9)	Notify QS of any disagreement with valuation	14 days from QS s notification	Contractor

Condition	Subject	Limit	Imposed on
42(12)	Deliver daywork vouchers to QS for valuation	By end of week following that in which work is done	Contractor
43(2)	Submit information to QS for valuation of other Instructions	28 days of complying with Instructions	Contractor
43(2)	Notify Contractor of amount determined	28 days of receipt of information	QS
43(3)	Notify QS of disagreement with amount determined	14 days of receipt	Contractor
46(3)	Provide details of expenses incurred as a result of prolongation and disruption	56 days of incurring expense	Contractor
46(5)	Notify Contractor of decision on expenses claimed	28 days from receipt of information referred to in Condition 46(3)(b)	QS
48A	Deliver retention payment bond	28 days of acceptance of tender	Contractor
48B(1)	Deliver mobilisation payment bond	28 days of acceptance of tender	Contractor
48B(1)	Pay mobilisation payment	14 days of delivery of mobilisation payment bond	Employer
49(2)	Send copy of draft final account to Contractor	6 months from certified completion	QS
49(2)	Notify agreement or disagreement with draft final account	3 months from receipt of draft final account	Contractor
50(2)	Issue certificates for payment	28 days from commencement of Works, and then monthly	PM
50(3)	Issue certificates for payment	21 days from Contractor's claim	PM
50(6)	Pay certified sum	30 days from certification and invoicing	Employer
50A(1)	Notify amount and basis of each payment	5 days after due date for payment	Paying party
50A(2)	Notice of intention to withhold payment	7 days before due date for payment	Party serving notice

Condition	Subject	Limit	Imposed on
52	Notice of intention to suspend performance	7 days before suspension	Party serving notice
56(1)	Warning notice of default	14 days from notice	Contractor
56(4)	After determination of the Contract, directions may be given in relation to performance or completion	Not later than 3 months from notice of determination or Date for Completion, whichever is sooner	Employer
58(1)	Warning notice of default	14 days from notice	Employer
59(3)	Appointment of adjudicator and named substitute adjudicator	28 days from acceptance of tender	Both parties
59(4)	Submission of representations to the adjudicator	Not later than 7 days from receipt of notice of referral	Recipient
59(5)	Notification of decision to PM, QS, Employer and Contractor	Not earlier than 10 and not later than 28 days from receipt of notice of referral	Adjudicator
60(1)	Notice of arbitration	56 days after notification of adjudicator's decision	Party serving notice
60(2)(c)	Arbitrator's timetable for delivery of points of claim, etc.	Not (without consent) to exceed 6 months from date of preliminary meeting	Arbitrator
60(2)(e)	Award to be made by arbitrator	Not later than 3 months from end of period in Condition 60(2)(c)	Arbitrator
66	Deliver performance bond	21 days from acceptance of tender	Contractor
67	Deliver parent company guarantee	21 days from acceptance of tender	Contractor

NOTE: Printed time limits can be extended by agreement within the terms of Condition 1(4). Days means calendar days - see Condition 1(1).

ALPHABETICAL INDEX

Item	Condition
I	
Indemnity against damage	19
Injury, personal	19
Insolvency of nominated subcontractors or suppliers	63A
Instructions	
non-compliance with	53
PM s	40
Insurance, general	8
Insurance, professional indemnity	8A
Intellectual property rights	12
L	
Liquidated damages	55
Loss of profits	19
M	
Maintenance Period	1,21,49
Measurement	18
Milestone	1,48
Milestone Payment Chart	1,48
Mobilisation payment	48B
N	
Nominated subcontractors or suppliers	63,63A
Nomination	63
Notices	1
Notices, statutory	11
Nuisance	14
O	
Occupier's rules and regulations	22
Official secrets	29
Old materials, credits for	48
Omissions	40
Other works	65
P	
Parent company guarantee 67	
Passes 27	
Payment	
certification of	50
mobilisation	48B
suspension for non-payment	52
Things off-site, for	48C
withholding	50A
Performance bond	66
Personal injury	19
Photographs	28
Planning Supervisor, definition of	1

Item	Condition
Pollution	14
Possession	
early	37
of the Site	34
Principal Contractor, definition of	1
Prime Cost items	63
Professional indemnity insurance	8A
Profits, loss of	19
Programme	1,33
Progress meetings	35
Project Manager (PM)	
definition of	1
Instructions of	40
non-compliance with Instructions	53
valuation of PM s Instructions	41,42,43
Prolongation and disruption	46
Protection of Works	13
Provisional sums	64
Q	
Quality	31
Quantity Surveyor (QS), definition of	1
R	
Racial discrimination	23
Rates, Schedule of	1
Records	25
Recovery of sums	51
Removal of Things from Site	30,40
Re-nomination	63
Replacement of employees	6
Representatives	4
Retention	48,48A,48C,49
Retention payment bond	48A
Resident Engineer	4
Returns	15
Risks, Accepted	1,8,19,36
S	
Savings, cost	38
Schedule of Rates	1
Secrets, official	29
Section, definition of	1
Setting out	9
Sex discrimination	23
Site	
admittance to	26
definition of	1
possession of	34
Specification, definition of	1
Stage Payment Chart	1,48

Item	Condition
Statutory notices	11
Strikes and industrial action	36
Subletting	62
Subcontractors and suppliers	62,63,63A
Subcontracts, determination of	62
Suspension for non-payment	52

T

Teamworking	1A
Things	
conformity	31
for/not for incorporation	1
off-Site, payment for	48C
removal of	30,40
Time, extensions of	36

U

Unforeseeable Ground Conditions	1,7

V

Valuation	
of advances on account	48
of Instructions-principles	41
of Variation Instructions	42
of other Instructions	43
by measurement	18
VAT	45
Variations	1,40
Vesting	30

W

Warranties, collateral	68
Works	
certifying	39
conditions affecting	7
covering	17
damage to	19
definition of	1
design of	10
insurance of	8
other	65
protection of	13
quality of	31

GC/WORKS/1 WITHOUT QUANTITIES (1998)

ABSTRACT OF PARTICULARS AND ADDENDUM

ABSTRACT OF PARTICULARS

Works:

Site:

Condition 1(1) (Definitions, etc.) Employer

The Employer shall be

of

Conditions 1(1) (Definitions, etc.): Project Manager, and 4(1) (Delegations and representatives)

The Project Manager shall be

*of/whose registered office is at

who shall act generally on behalf of the Employer in carrying out those duties described in the Contract, subject to the following excluded matters:

In relation to such excluded matters, the person or persons authorised to act for the Employer are:

The Planning Supervisor shall be

*of/whose registered office is at

*All the CDM Regulations apply/Only Regulations 7 and 13 of the CDM Regulations apply.

Condition 8 (Insurance)

The minimum amount insured in respect of employer's liability referred to in Condition 8(2) shall not be £10,000,000, but shall be £

In respect of Condition 8(3), Alternative *A/B/C is required.

The percentage for professional fees referred to in *Condition 8(3)(a)/the Summary of Essential Insurance Requirements shall not be 15%, but shall be   %.

The minimum amount insured in respect of public liability referred to in *Condition 8(3)(b)/the Summary of Essential Insurance Requirements shall be £   for any one occurrence or series of occurrences arising out of one event.

The forms of certificates *and Summary of Essential Insurance Requirements referred to in Condition 8 are appended.

Condition 8A (Professional indemnity insurance for design)

Condition 8A *shall/shall not apply.

*The amount of professional indemnity insurance required is £

*The period during which professional indemnity insurance is required shall end years after certification under Condition 39 (Certifying completion) of the completion of the Works or the last Section thereof in respect of which completion is certified, or the determination of the Contract for any reason whatsoever, including (without limitation) breach by the Employer, whichever is the earlier, and not 12 years after such certification.

Condition 10 (Design)

*The Contractor (or a subcontractor) is required to undertake the design of the following part or parts of the Works:

Alternative A is required*/

Alternative B is required*/

Tenderers are required to tender alternative prices respectively for Alternatives A and B*.

The drawings, design documents and design information shall be supplied in copies, and not 2 copies.

Condition 21 (Defects in Maintenance Periods)

Other than for the services listed below, the Maintenance Period for the Works (or each Section where completion is required in Sections) shall be   months and shall apply from the day after that on which the Works (or each Section) are completed as certified by the PM.

The Maintenance Period for each of the following services, which shall apply from the day after that on which the Works (or each Section where completion of the Works is required in Sections) are completed as certified by the PM, shall be:

Service	Period
	months.
	months.
	months.

Condition 22 (Occupier's rules and regulations)

Condition 22 *shall/shall not apply.

*The occupier's rules and regulations are appended.

Condition 27 (Passes)

Condition 27 *shall/shall not apply.

Condition 28 (Photographs)

Condition 28 *shall/shall not apply.

Condition 34 (Commencement and completion)

Period within which notice of date of possession to be given: Days of the acceptance of the tender. (In the absence of notice the Contractor may take possession 14 Days after acceptance of tender).

Period for completion of the Works shall be the Day after the expiration of a period of *weeks/months from the *date so notified/acceptance of the tender.

Periods for completion of the Sections shall be the Day after the expiration of the period set out below opposite each Section from the *date so notified/acceptance of the tender:

Section	Period
1	*weeks/months.
2	*weeks/months.
3	*weeks/months.

Condition 38A (Bonuses)

Condition 38A *shall/shall not apply

*The rate of bonus for early completion shall be: £   per Day.

*The rates of bonus for early completion of each Section shall be:

Section		
1	£	per Day.
2	£	per Day.
3	£	per Day.

Condition 47 (Finance charges)

The rate at which finance charges shall be payable shall be   % over the rate charged during the relevant period by the Bank of England for lending money to the clearing banks.

Condition 48 (Advances on account)

Alternative A (Stage Payment Chart)*/

Alternative B (Milestone Payment Chart)*/

Alternative C (Valuation)*

is required.

*If Alternative A (Stage Payment Chart) applies, examples of the prescribed forms of chart and chart banding calculation sheet are appended.

*If Alternative B (Milestone Payment Chart) applies, the prescribed form of chart is appended.

*If Alternative B (Milestone Payment Chart) applies, Contractor's applications under Condition 48(2)(a) shall also include all amounts due under Condition 48(2)(b)-(h).

Condition 48A (Retention payment bond)

*The Employer will pay the Contractor advances on account without deduction of retention, provided that the amount of retention so foregone by the Employer shall not exceed the following Retention Payment: £

*The prescribed form of retention payment bond is appended.

Condition 48B (Mobilisation payment)

Condition 48B *shall/shall not apply.

*Mobilisation payment: % of the Contract Sum.

*Percentage for recovery of mobilisation payment: %.

*The prescribed form of mobilisation payment bond is appended.

Condition 48C (Payment for Things off-Site) (not applicable in Scotland)

Condition 48C *shall/shall not apply.

If applicable, Condition 48C shall apply compulsorily in respect of the *whole of the Works/only in respect of the following part or parts of the Works:

and shall *apply voluntarily/shall not apply in respect of the rest of the Works.

Condition 50 (Certifying payments)

The prescribed form of certificate is appended.

Condition 55 (Liquidated damages)

Damages for delay shall be: £   per Day.

Damages for delay in respect of each Section shall be:

Section

1	£	per Day.
2	£	per Day.
3	£	per Day.

*Condition 58 (Determination by Contractor)

*The period of suspension referred to in Condition 58(3)(e) shall not be 182 Days, but shall be   Days.

*Condition 58A (Determination following suspension of Works)

*The period of suspension referred to in Condition 58A(1) shall not be 182 Days, but shall be   Days.

**Condition 59 (Adjudication)

The adjudicator shall be

of

or, if he is deceased or unwilling or unable to act, or is not or ceases to be independent of the Employer, the Contractor, the PM and the QS,

of

or, if he is deceased or unwilling or unable to act, or is not or ceases to be independent of the Employer, the Contractor, the PM and the QS; such other person as the Employer and the Contractor choose by mutual agreement in writing or, failing such agreement, such other person as may be chosen by the President or a Vice President of the Chartered Institute of Arbitrators (or, where the Contract is a Scottish contract, by the Chairman or a Vice Chairman of the Chartered Institute of Arbitrators (Arbiters) (Scottish Branch)) at the request of either the Employer or the Contractor.

The prescribed form of adjudicator's appointment is appended.

**Condition 60 (Arbitration and choice of law)

The arbitrator shall be

of

or, if he is deceased or unwilling or unable to act, or is not or ceases to be independent of the Employer, the Contractor, the PM and the QS,

of

or, if he is deceased or unwilling or unable to act, or is not or ceases to be independent of the Employer, the Contractor, the PM and the QS; such other person as the Employer and the Contractor choose by mutual agreement in writing or, failing such agreement, such other person as may be chosen by the President or a Vice President of the Chartered Institute of Arbitrators (or, where the Contract is a Scottish contract, by the Chairman or a Vice Chairman of the Chartered Institute of

Arbitrators (Arbiters) (Scottish Branch)) at the request of either the Employer or the Contractor.

Condition 61 (Assignment)

The Employer may assign or transfer the benefit of the Contract, or any part, share or interest under it, *either before or after/only after certification by the PM under Condition 39 (Certifying completion) of the completion of the Works or the last Section thereof in respect of which completion is certified, or the determination of the Contract for any reason whatsoever, including (without limitation) breach by the Employer, whichever is the earlier.

Condition 62 (Subletting)

*The following subcontract terms and conditions are prescribed, or the following model form of sub-contract:

Condition 63A (Insolvency of nominated subcontractors or suppliers)

Condition 63A *shall/shall not apply.

Condition 66 (Performance bond)

Condition 66 *shall/shall not apply.

*The performance bond shall be in an amount of   % of the Contract Sum, and not 10%.

*The prescribed form of performance bond is appended.

Condition 67 (Parent company guarantee)

Condition 67 *shall/shall not apply.

*The prescribed form of parent company guarantee is appended.

Condition 68 (Collateral warranties)

Condition 68 *shall/shall not apply.

*The prescribed forms of collateral warranty and parent company guarantee are appended.

*The period during which subcontractors' professional indemnity insurance is required shall be the same as that required for the Contractor's professional indemnity insurance under Condition 8A (Professional indemnity insurance for design)/   years after certification under the Main Contract of completion of the Works or the last Section thereof in respect of which completion is certified, or the determination of the Main Contract for any reason whatsoever, including (without limitation) breach by the Employer or the Contractor, whichever is the earlier.

Contract Agreement

The prescribed form of Contract Agreement is appended.

***Supplementary Conditions and Annexes

The following Supplementary Conditions and Annexes (if any) are incorporated into the Conditions of Contract, and shall prevail over the other Conditions of Contract:

*Delete inapplicable items.

**The same adjudicators and arbitrators should be named in all the Employer s contracts relating to the project, whether with contractors, consultants or others.*

***It is recommended that any printed Conditions affected by Supplementary Conditions should be amended and initialled by both parties.*

ADDENDUM TO ABSTRACT OF PARTICULARS

Schedule of Design Information

Information on items in Condition 46(2)(a) and (b) listed below is not yet available but will be provided by the PM within the periods indicated below. Items not listed will be provided in time to meet the Contractor's reasonable requirements where these have been notified in reasonable time.

Item	**Weeks

Directions or Instructions by the Employer or the PM under Condition 46(2)(c) will be given within the periods indicated below.

Nominated Subcontract	**Weeks

* Delete inapplicable items.

** The weeks are counted from the date upon which the Contractor may take possession as notified to him by the Employer, or from the acceptance of the tender, whichever is provided by the Abstract of Particulars (see Condition 34(1) (Commencement and completion)).

GC/WORKS/1 WITHOUT QUANTITIES (1998)

INVITATION TO TENDER AND SCHEDULE OF DRAWINGS

INVITATION TO TENDER

Works:

Site:

1   You are invited on behalf of (the Employer) to tender, upon the basis of GC/Works/1 Without Quantities (1998), for the Works described in the following enclosed documents:

(a)   Abstract of Particulars and Addendum;

(b)   Supplementary Conditions and Annexes (if any) referred to in the Abstract of Particulars;

(c)   Specification;

(d)   Drawings listed in the attached Schedule of Drawings;

(e)   Schedule of Rates (including Stage or Milestone Payment Chart (if used));

(f)   Outline Health and Safety Plan; and

(g)   Other documents as listed below:

Chart banding calculation sheet*

2   Your tender should be submitted on the form of Tender and Tender Price Form also enclosed. Any obvious errors in pricing or errors in arithmetic will be dealt with as stated in the form of Tender.

3   You are required to keep your tender confidential and not divulge to anyone, even approximately, what your tender price is or will be. The sole exception to this is information you may have to give to your insurance company, or broker, in order to compile your tender, but you must stress to them that this information is given in strict confidence.

4   You must not make any arrangements with anyone else about whether or not they should tender, or about their or your tender prices or terms and conditions. You may however, obtain any necessary subcontract quotations.

5   No tendering expenses will be reimbursed by the Employer.

6   Tenders received late will not be considered unless due to genuine postal delays. If the tender is qualified it may be set aside, or you may be required to withdraw the qualification without amending your offer. Any proposals for alternatives to the specified requirements should be submitted by way of a separate, unqualified, bid after checking with the Employer on the procedure to follow.

7   The Employer does not bind himself to accept the lowest, or any, tender.

8   Your form of Tender should be submitted in a sealed envelope prominently marked:

FORM OF TENDER FOR WORKS:

SITE:

The envelopes should bear no external indication of the identity of the tenderer.

8   Tenders must be completed and returned by   a.m./p.m on
     to:

SIGNED by

for and on behalf of the Employer

Tel:

Fax:

Telex:

Date:

A chart banding calculation sheet will only be required if a Stage Payment Chart is used.

SCHEDULE OF DRAWINGS

Drawings prepared by

Discipline

Drawing No. & Revision No. (if any)	Drawing Title	Date

GC/WORKS/1 WITHOUT QUANTITIES (1998)

TENDER AND TENDER PRICE FORM

TENDER

Works:

Site:

To be returned by   a.m./p.m. on   to

of

1   We have examined GC/Works/1 Without Quantities (1998), and the following documents:

   (a) Abstract of Particulars and Addendum;

   (b) Supplementary Conditions and Annexes (if any) referred to in the Abstract of Particulars;

   (c) Specification;

   (d) Drawings listed in the Schedule of Drawings;

   (e) Schedule of Rates (including Stage or Milestone Payment Chart (if used));

   (f) Outline Health and Safety Plan (and confirm that we will provide a statement and details of how we plan to implement and develop it, together with details to establish our competence and resources to comply with the requirements and prohibitions imposed upon us relative to health and safety in the execution and/or management of the Works); and

   (g) Other documents as listed below:

2   We enclose for your approval our Programme, our priced Schedule of Rates, and the other enclosed documents, which shall be deemed to form part of our tender, listed below:

   Chart banding calculation sheet, duly completed*.

3   We have obeyed the rules about confidentiality of tenders and will continue to do so as long as they apply.

4   We undertake to satisfy the Employer that the prices in the Schedule of Rates are fair, and should reasonably be used to value Variation Instructions.

5   We agree that, should errors in pricing or errors in arithmetic be discovered in any schedules of rates submitted by us during consideration of this offer, we will, in addition to the chance to confirm the offer as tendered despite the errors, be afforded the opportunity of *either* - withdrawing it -or- correcting it with appropriate explanations.**

6   Subject to and in accordance with paragraphs 3 to 5 above and the terms and conditions contained or referred to in the documents listed in paragraphs 1 and 2, we offer to execute the Works referred to in the said documents in consideration of payment by the Employer of the sum shown in our accompanying Tender Price Form, which shall be deemed to form part of our

tender, plus reimbursement by the Employer of Value Added Tax in accordance with Condition 45 (VAT).

7   (only applicable if Abstract of Particulars states that Condition 8A (Professional indemnity insurance for design) shall apply)

Our professional indemnity insurance is at least that required by the Abstract of Particulars. Details of the insurance are as follows:

Insurers:

Policy No.:

Renewal Date:

8   (only applicable if Abstract of Particulars states that payment will be made without deduction of retention under Condition 48A (Retention payment bond) and/or that a mobilisation payment will be made under Condition 48B (Mobilisation payment) and/or that Condition 66 (Performance bond) shall apply)

Our surety/sureties will be   Limited/PLC, whose registered office is at

9   (only applicable if Abstract of Particulars states that Condition 67 (Parent company guarantee) shall apply)

Our ultimate holding company (if any) is   Limited/PLC, (No.), whose registered office is at

10   We undertake, within 21 days of being so required by the Employer, to enter into a Contract Agreement with the Employer in duplicate (as a Deed, if so required by the Employer) in the form included in the said documents.

11   We agree that differences or questions arising out of or relating to the Contract shall be resolved in accordance with Conditions 59 (Adjudication) and 60 (Arbitration and choice of law) of the General Conditions.

SIGNED by

for and on behalf of

Tel:

Fax:

Telex:

Date:

A chart banding calculation sheet will only be required if a Stage Payment Chart is used.

** *Employer to delete which of the 2 options will not be offered before issuing tender documents.*

TENDER PRICE FORM

Works:

Site:

To be returned by   a.m./p.m. on   to

of

The sum referred to in our accompanying form of Tender is   pounds (£   )*/

The sum referred to in our accompanying form of Tender is   pounds (£   ) if Condition 10 (Design) (Alternative A) applies, or   pounds (£   ) if Condition 10 (Design) (Alternative B) applies*.

We enclose the Milestone Payment Chart, duly completed*.

SIGNED by

for and on behalf of

Tel:

Fax:

Telex:

Date:

*Employer to delete inapplicable items before issuing tender documents.

GC/WORKS/1 WITHOUT QUANTITIES (1998)

CONTRACT AGREEMENT (ENGLAND, WALES & NORTHERN IRELAND)

Works:

Site:

THIS AGREEMENT made the   day of -

BETWEEN:

(1)

of

('the Employer'); and

(2)

[of] OR [whose registered office is at]

('the Contractor');

INCORPORATES the General Conditions of Contract for Building & Civil Engineering Major Works GC/Works/1 Without Quantities (1998), and the other documents comprising the Contract as defined in the General Conditions, copies of all of which are annexed or have been signed for identification purposes by or on behalf of the Employer and the Contractor. Any disputes, differences or questions arising out of or relating to the Contract shall be resolved in accordance with Conditions 59 (Adjudication) and 60 (Arbitration and choice of law) of the General Conditions.

IN WITNESS whereof the parties have executed this Agreement in duplicate on the date first stated above.

GC/WORKS/1 WITHOUT QUANTITIES (1998)

CONTRACT AGREEMENT (SCOTLAND)

Works:

Site:

AGREEMENT

BETWEEN:

(1)

of ('the Employer'); and

(2)

[of] OR [whose registered office is at]

('the Contractor');

WHEREAS the Employer has agreed to employ the Contractor to undertake the execution and completion of the Works as defined in the Conditions hereinafter specified, and the Contractor has agreed to do so;

THEREFORE the Employer and the Contractor HAVE AGREED and DO HEREBY AGREE as follows:

1.   The Works shall be completed in accordance with and the rights and duties of the Employer and the Contractor shall be regulated by:

    1.1   the General Conditions of Contract for Building & Civil Engineering Major Works GC/Works/1 Without Quantities (1998) ('the Conditions');

    1.2   the Abstract of Particulars annexed hereto;

    1.3   *the Supplementary Conditions referred to in the Abstract of Particulars and annexed hereto;

    1.4   *the Annexe(s) referred to in the Abstract of Particulars and annexed hereto;

    1.5   the Specification annexed hereto;

    1.6   the Drawings listed in the Schedule of Drawings annexed and signed as relative hereto;

    1.7   the Schedule of Rates annexed hereto;

    1.8   the Programme annexed hereto;

    1.9   the Contractor's tender dated annexed hereto; and

    1.10 the Employer's letter of acceptance dated   annexed hereto;

    all of which are held to be incorporated in and form part of the Contract.

2.   Any disputes, differences or questions arising out of or relating to the Contract shall be resolved in accordance with Conditions 59 (Adjudication) and 60 (Arbitration and choice of law) of the Conditions.

3   Both parties consent to registration hereof for preservation and execution.

IN WITNESS WHEREOF these presents typewritten on this and the preceding [   ] page[s] are executed as follows:

Delete inapplicable items.

ACKNOWLEDGEMENTS

The Government agency responsible for this document is -

Property Advisers to the Civil Estate (PACE)
Trevelyan House
Great Peter Street
London SW1P 2BY

Tel: (0171) 271 2833 Fax: (0171) 271 2715

The development of this document was directed and greatly aided by a sub-group of the PACE Joint Users' Group (JUG) of Government Departments, consisting of -

Bruce Perry	PACE Central Advice Unit (Chairman)
Robert Boyd	Department of the Environment for Northern Ireland Construction Service
Charles Branch	Department of Trade and Industry
Ian Campbell	The Scottish Office
Mike Frazer	Benefits Agency Estates
John Garnett	Consultant to the Ministry of Agriculture, Fisheries and Food
Jeff Hogg	Lord Chancellor's Department
Jay Jayasundara	HM Treasury (Procurement Practice & Development Group, formerly Central Unit on Procurement)
Lynne Jones	The Buying Agency
Graham Mason	Prison Service
Deric McTier	Department of the Environment for Northern Ireland Construction Service
Robert Pilling	PACE Central Advice Unit
David Reid	Home Office
Bill Robinson	Employment Service
Richard Whittaker	English Heritage
Les Wilson	British Nuclear Fuels plc
David Woolger	Inland Revenue
Neil Wright	Benefits Agency Estates
Harry Yeabsley	Ministry of Defence

The substantive drafting and legal advice and assistance necessary for the preparation of this document was principally provided by Andrew Pike of -

Pinsent Curtis
Solicitors
3 Colmore Circus
Birmingham B4 6BH

Advice on Scots law was provided by David Henderson of -

MacRoberts
Solicitors
152 Bath Street
Glasgow G2 4TB

Advice on Northern Ireland law was provided by -

Departmental Solicitor's Office
Department of Finance & Personnel
Victoria Hall
12 May Street
Belfast BT1 4NL

Advice on risk and insurance matters was provided by Robin Keeling of -

Aon Group Limited
Construction Division
15 Minories
London EC3N 1NJ

Construction consultancy advice was provided by Brendan Murphy of -

Tarmac Services
The Lansdowne Building
Lansdowne Road
Croydon CR0 2BX

Printed in the United Kingdom for The Stationery Office
J0042757 3/98 C25 10170